Palgrave Series in Asia and Pacific Studies

The Asia and Pacific regions, with a population of nearly three billion people, are of critical importance to global observers, academics, and citizenry due to their rising influence in the global political economy as well as traditional and nontraditional security issues. Any changes to the domestic and regional political, social, economic, and environmental systems will inevitably have great impacts on global security and governance structures. At the same time, Asia and the Pacific have also emerged as a globally influential, trend-setting force in a range of cultural arenas. The remit of this book series is broadly defined, in terms of topics and academic disciplines. We invite research monographs on a wide range of topics focused on Asia and the Pacific. In addition, the series is also interested in manuscripts pertaining to pedagogies and research methods, for both undergraduate and postgraduate levels. Published by Palgrave Macmillan, in collaboration with the Institute of Asia and Pacific Studies, UNNC.

More information about this series at
http://www.palgrave.com/gp/series/14665

Ali Cheshmehzangi • Chris Butters
Editors

Designing Cooler Cities

Energy, Cooling and Urban Form: The Asian Perspective

palgrave
macmillan

Editors
Ali Cheshmehzangi
University of Nottingham
Ningbo, Zhejiang
China

Chris Butters
Warwick University
Coventry, United Kingdom

Palgrave Series in Asia and Pacific Studies
ISBN 978-981-10-6637-5 ISBN 978-981-10-6638-2 (eBook)
https://doi.org/10.1007/978-981-10-6638-2

Library of Congress Control Number: 2017955269

Cover illustration: Ali Cheshmehzangi

Printed on acid-free paper

This Palgrave Macmillan imprint is published by Springer Nature
The registered company is Springer Nature Singapore Pte Ltd.
The registered company address is: 152 Beach Road, #21-01/04 Gateway East, Singapore 189721, Singapore

We would like to dedicate this book to our friend and colleague, Dr. Terry Thomas, *and to the welfare of the millions increasingly exposed to urban heat island effects.*

For Ayra—*always aim high!*

Foreword

Keeping cool, in hot climates, has in recent decades become a matter for public concern and action affecting whole urban populations. Air-conditioning has extended from the rich to the middle classes in hot-climate cities and will soon be accessible to the fairly poor. City planners have expanded a two-centuries-old interest in the 'urban heat island' effect to include detailed thermal modelling of blocks and districts. Urban cooling has created a massive new demand for energy in rapidly developing countries and cooling energy is set to globally overtake that required for winter heating in the cold climates. The recent field of district cooling is beginning to replace building-by-building air-conditioning. At last, tropical architecture is returning to its traditions of climatically responsive solutions, though now with a mixture of passive design and mechanised cooling technologies.

Several forces are working to make hot cities even hotter, including global warming, hardscape and a steady growth in the release of heat from humans, vehicles and (ironically) from cooling plant itself. Rises in peak urban temperatures of up to 6 °C by the year 2100 are being forecast. The energy and greenhouse gas emissions associated with cooling are very serious, but so are the human consequences. Given rising heat stress and indeed mortality from high temperatures and heat waves, this is also a matter of public health policy. Cooling is even provoking discussion of rich-poor thermal inequity. Lower productivity is yet another consequence of the urban heat island effect.

This timely book has arisen in part from a 3-year, 6-partner study of 'Energy and Low-income Tropical Housing', funded by the UK government.

Edited by an architect and a town planner, it addresses urban cooling on three levels: individual buildings—urban neighbourhoods—whole cities. Its contributors address the concerns noted above and explore how cooling needs can at once be reduced, and supplied in ways that avoid the unwanted positive feedback of releasing ever more 'reject heat' into the streets and atmosphere of large cities. The Asian focus in the case studies reflects the expectation that soon most of the world's megacities will be in tropical Asia.

T.H. Thomas
Director of the EPSRC-funded ELITH Research programme
School of Engineering, Warwick University, UK

Acknowledgement

This work stems partly from research activities carried out within the ELITH program funded by EPSRC, UK (2013–2016). The editors wish to thankfully acknowledge their funding; views expressed are not necessarily those of EPSRC or the funding ministries DfID and DECC. The ELITH web site is at: http://www2.warwick.ac.uk/fac/sci/eng/elith/

About This Book

Following a very brief discussion of cities, energy and cooling, we present examples that illustrate both research and practice in the field of cooling in hot climate cities. Whilst building on extensive scientific sourcing, we avoid detailed technical information, which is available in the specialised literature. The chapters and case studies in this book represent a deliberate mix of design, research, planning and policy approaches. Whilst our readers may have particular interest in one or two of these fields, we emphasise the need for interdisciplinary understanding and processes. Policy makers and urban planners often have little contact with the latest research, and researchers for their part often have limited interaction with the real world faced by the decision makers.

Conventional city policy and planning seldom consider which is the *optimal level* for sustainable solutions. This distinction of levels is vital. Is it most advantageous or cheapest to solve problems at the level of individual buildings, at the larger scale of an urban district, or at the level of the overall city energy system? One may broadly discern three 'levels' of action: termed here the micro, meso and macro levels. The micro level pertains to the scale of individual buildings; the meso or intermediate level to larger developments, such as a housing estate or office precinct; and the macro level to whole urban districts or cities. This book provides sections addressing each of these levels, with examples illustrating concrete solutions, scientific research, and overarching planning or policy questions.

On the micro level we present examples of individual buildings that reduce environmental impacts including the need for cooling, and a chapter discussing research and policy processes on a national basis towards energy efficient buildings. Our meso level section includes a research study on creating cooler city microclimates using vegetation, an example of a climate-adapted residential neighbourhood, and a concept design for a dense urban development that aims to integrate all three aspects of building design, planning layout and energy supply system. On the macro level, finally, we again present examples from different perspectives: climate-adapted sustainable city design, scientific research into cooler city layouts, and macro-scale energy solutions for hot climate cities with their policy implications. The interconnections between these three levels are emphasised throughout the book.

A good cross-section of Asian countries is represented: China, Thailand, India, Sri Lanka, Malaysia and Singapore. Whilst the examples are selected from Asian hot climate cities, they are largely relevant for other hot climate contexts too. For a comparative perspective, we also draw on European state-of-the-art experience. Vernacular traditions are noted, for they contain evergreen wisdom, which we can, and indeed should, apply in new ways. A big challenge lies in identifying solutions that suit local cultural and governance conditions; technically excellent solutions in one country may be totally inapplicable in another. We face the questions: how relevant is recent temperate climate research for hot climate cities? What lessons are transferable? What particular opportunities can be found in hot climate developing countries? In our concluding chapters we, therefore, add reflections on types of city, on the dynamics and barriers to sustainable solutions, and a few key recommendations.

Contents

Editors and Contributors

Notes on Editors

Ali Cheshmehzangi holds a PhD in Architecture and Urban Design (Nottingham), a Masters in Urban Design (Nottingham), a Graduate Certificate in Professional Studies in Architecture (Leeds), and a Bachelor degree in Architecture (Leeds). He is an urbanist and urban designer by profession. Ali is currently Associate Professor of Architecture and Urban Design at The University of Nottingham Ningbo China and a founding director of the International Network for Urban and Rural Research (INURR). He was the China-lead member of the ELITH programme. Ali has previously worked in several UK universities and practices, and has worked on several practice and research projects on eco-cities in China (Cao Feidian eco-city, Meixi Lake eco-city, Chongming Island, etc.), low-carbon town planning, urban modelling of residential neighbourhoods in several countries, green infrastructure of cities, toolkit for resilient cities (with Arup and Siemens), sponge city development and green development in Ningbo City, and other projects related to eco/green/smart city developments in various contexts. He has developed a comprehensive planning toolkit, called 'Integrated Assessment of City Enhancement (iACE)'. More recently, he has developed a new theme of 'Eco Fusion' and has been working on a comparative study of eco and smart transitions between the European and Chinese cities.

Chris Butters, a Norwegian citizen, holds a BA in Literature (Stellenbosch), a Masters in Architecture DPLG (Montpellier) and a Diploma in Energy Planning (Oslo). He has worked with architecture, energy, ecology and sustainable planning for 35 years, as practising architect, author, lecturer and international consultant. He is the author or co-author of eight books. Formerly director of Norwegian Architects for Sustainable Development and senior consultant at the Ideas Bank Foundation. More recently, Chris has been researcher at Warwick University, UK, on the ELITH programme noted in this book. He has been guest professor at the Oslo School of Architecture and at St. Johns University, Minnesota, USA. He leads a post-graduate program on Energy and Sustainable Development at the Centre for Development and the Environment, Oslo University. Chris worked for 10 years in Bhutan as architect and project coordinator for schools and a hospital, as well as doing Tibetological research. He has worked in countries including France, Norway, UK and Bhutan and is a co-founder of the GAIA group in Norway (1982) and GAIA International (1992), a pioneering network for sustainable building and 'deep-green' solutions. He is also the creator of the widely used evaluation tool, the Sustainability Value Map.

Notes on Contributors

Amir Cheshmehzangi holds a Diploma in Architecture part II from The University of Nottingham, UK. He is an architectural and urban designer and has held several practice positions. He is currently Managing Director and Architectural Designer at AVA Consultants International Ltd., Derby. He has practical and research experience in sustainable architecture, urban planning and development. He has taught architecture studios at The University of Nottingham and has led various workshops. He has completed architectural and interior design projects in residential planning, and refurbishment projects in the UK, and several residential design projects internationally.

Shiuh-Shen Chien obtained his PhD from the London School of Economics and Political Science and is currently Professor of Development Geography and Urban and Environmental Studies, at the National Taiwan University. His research interests cover urban policy, mobility and transnational learning, the political economy of local and regional development, smart-eco urbanism, climate change and environmental governance, and international development of the global South. His ongoing research projects include 'Theorization of China's Urban Speed Machine' and 'Climate Engineering and Weather Modification in China'.

Ayotunde Dawodu is a PhD candidate at the International Doctoral Innovation Centre (IDIC), and the Department of Architecture and Built Environment, University of Nottingham Ningbo China. He holds a BSc degree in Mechanical Engineering (Lagos) and MSc in Sustainable Energy Engineering (Nottingham). His Research is on the built environment and is focused on developing a Neighbourhood Sustainability Assessment Tool (NSAT) for Lagos, Nigeria. He has worked and published on themes of 'Sustainability Indicators', 'Water and Waste Management', 'Sustainable Energy', and 'Governance and Management of City Communities'.

Yueyang He has been a doctoral researcher at the Department of Architecture of the National University of Singapore (NUS) since 2014. He has a bachelor degree in urban planning from the Sun Yat-sen University and a Masters degree in urban environmental planning from the Chinese University of Hong Kong. He is currently working on a doctorate which focuses on high-density urban ventilation and urban breezeway patterns in Singapore and other Asian cities.

Dong-Li Hong is a research assistant in the Department of Geography at the National Taiwan University. He received his MSc degree at the Institution of Building and Planning, NTU, and will enter the School of Geography and the Environment at University of Oxford, UK for the DPhil from late 2017. Dong-Li is interested in environmental governance, bio-politics and the political ecology of the Anthropocene, especially in the Asian context. His current work focuses on experimental wilderness conservation projects in China, in which weather modification has been applied as a means of ecological restoration.

Yun Hye Hwang is Assistant Professor in the Department of Architecture of the National University of Singapore. With an MLA degree from Seoul National University and Graduate School of Design, Harvard, USA, she has been teaching in landscape design studios and research dissertation since 2009. Her teaching and professional activities focus on emerging demands of landscapes in the Asian equatorial urban context by exploring dynamic and sustainable landscape management versus highly manicured green spaces, and developing design strategies to counter anthropogenic ecological challenges in fast growing megacities.

Danny Tam Hong Khai is the General Manager at Shinryo Corporation's design and engineering centre for District Energy projects based in Malaysia. He holds a Bachelor degree in Mechanical Engineering from the National University of Singapore. He has been involved in the District Energy industry since 1998, and has worked in the turnkey construction of District Cooling and Cogeneration plants in Malaysia, Thailand, Singapore and the UAE. Danny is an ardent advocate of District Energy, and regularly shares his industry experience at lectures/seminars and conferences organised by the Institution of Engineers Malaysia, ASHRAE, Fleming Gulf, Hong Kong Institution of Engineers.

Li Xuan Cherlyn Lim graduated from the National University of Singapore (NUS) in 2016 with a concurrent BA in Architecture and an MLA. She worked as a research assistant in NUS in 2015–2016 and is currently working in Ramboll studio Dreiseitl as a landscape architect.

Qin Jie Geraldine Lum graduated from the National University of Singapore (NUS) in 2011 with a concurrent BA in Architecture and an MLA. She worked as a research assistant in NUS in 2013–2014 and is currently working in the Gardens by the Bay development, Singapore as a landscape architect.

Adzuieen Nordin is senior lecturer in the Department of Mechanical Engineering at the Politeknik Ungku Omar of Malaysia. She holds a BSc in Mechanical Engineering (Malaysia), MEd Technical and Vocational (Malaysia), and a PhD degree from Universiti Teknologi Petronas. Since 2011, she leads research and innovation at the Center of Technology in Air-Conditioning and Refrigeration. Her research interests include energy systems, cogeneration plants and district cooling. She is a registered

ASHRAE member and is also a Graduate Member (Mechanical Engineering) of the Institution of Engineers, Malaysia (IEM).

Upendra Rajapaksha is senior lecturer in architectural design and architectural science at the Department of Architecture of University of Moratuwa, Sri Lanka. He holds a PhD degree in architecture specialising in sustainable design, from the University of Queensland, Australia; and since then, he is engaged in research and practice in green buildings, low energy architecture, carbon constrained buildings and environmental sustainability in the built environment. He has over 19 years of involvement as a practicing architect and 12 years in research and teaching architecture in Sri Lanka.

Pattana Rakkwamsuk is Assistant Professor in the School of Energy, Environment and Materials at King Mongkut's University of Technology Thonburi (KMUTT), Thailand. He was also part of the ELITH research programme. He holds a BSc degree in Physics (Bangsaen, Thailand), MSc in Energy Technology (Bangkok), MSc in Electrophysics (Washington DC), and a Doctorate in Electrophysics (Washington DC). Between 2006 and 2014 he was the Dean of the School of Energy, Environment and Materials at KMUTT, Thailand. His research focus is in the fields of energy conservation in buildings and industries, energy efficient building envelope materials, thin film technology, and electromagnetic theory.

Abel Tablada has been an Assistant Professor at the Department of Architecture of the National University of Singapore (NUS) since 2012. He has his Architecture degree from the Polytechnic University of Havana, Cuba, a Masters degree Magna Cum Laude and a PhD in Engineering from Katholieke Universiteit Leuven. His research and teaching interests include tropical net-zero/low-carbon architecture, urban morphology and natural ventilation, thermal comfort, building simulation and renewable resources. At present he is leading research on productive facades that integrate solar and farming systems. He recently published the book *Tropical Porosity: Wind-driven Design at Jurong Industrial and Residential Development.*

Linjun Xie is a PhD candidate in the Department of Architecture and Built Environment and School of International Studies at The University of Nottingham Ningbo China. Her doctoral work investigates the governance modes adopted in the experimental cities towards sustainability, which explores the transitional management in Chinese eco-city, low-carbon city and smart city developments. She holds a Masters Degree in Eco-cities (distinction) from Cardiff University, and bachelor degree in Urban Planning (Xiamen). Her current research revolves around urban sustainable development, including urban sustainability governance and policies, sustainability transitions, political ecology and social justice.

List of Figures

List of Tables

CHAPTER 1

Introduction

Ali Cheshmehzangi and Chris Butters

The Environmental Challenge of Our Cities

With rising affluence and rapid urbanisation, the energy and climatic, as well as, health impacts of cities are of increasing importance. Much of this intensive urban growth is in the hot climate developing countries, not least in Asia. The quest for 'sustainable' cities has led to innovative and successful solutions that offer better living quality along with greatly reduced environmental impacts. Due partly to a lack of resources, less has to date been done in developing countries and hot climates. The largest energy and climatic challenge in these cities is that of cooling.

Our urban environment determines indoor climate and wellbeing, as well, as outdoor comfort and public health. Whereas tackling the issue of heat in cities is our focus, sustainable development is essentially about quality of life, for both people and for the environment, both now and in the future. Seen as a whole—which it must be—the aim is a good balance between all three areas of ecology, economy and society. In the real world, however, economic and socio-political factors often weigh more heavily in

A. Cheshmehzangi (✉)
The University of Nottingham Ningbo China, Ningbo, China

C. Butters
University of Warwick, Coventry, UK

A. Cheshmehzangi, C. Butters (eds.), *Designing Cooler Cities*,
Palgrave Series in Asia and Pacific Studies,
https://doi.org/10.1007/978-981-10-6638-2_1

decision making than environment. This represents missed opportunities, where growing cities are locking themselves into poor health as well as huge future energy and climate burdens. Yet many sustainable solutions exist already, and do not even necessarily cost more. This applies equally in the field of cooling.

Hot-climate cities are increasingly problematic due to congestion, pollution and deteriorating microclimates, not least growing urban heat island (UHI) effects (Santamouris 2007). As is well known, inner city temperatures are commonly several degrees hotter than the surrounding countryside; leading to increased energy needs for cooling, as well as heat stress and indeed mortality. Global warming and rapid urbanisation both exacerbate overheating in cities. In this book, we focus on case studies from Asian hot climates because projections suggest they will be subjected to more land degradation, population displacement and economic disruption from climatic changes (including sea level rises) than any other part of the planet (Anthoff et al. 2006; USAID 2010). Our selected region is thus the most vulnerable to climate change and temperature increase. A megacity like Manila in the Philippines has already reached the danger level of 2 °C increase in its urban core—the increase to be avoided as part of the 2015 Paris climate agreement. Hence, we face a major challenge for both existing and new cities, and particularly for the urban poor who cannot afford better living environments or cooling amenities.

There is abundant research, as well as experience, about energy, transports, pollution, green buildings and other specific urban environmental topics. Much of it is addressed by specialists, one issue at a time. Our aim is to highlight the whole picture and address the issue of urban cooling in an interdisciplinary way not often found in practice but essential for sustainable outcomes. This book is, therefore, addressed to a wide readership of designers, urban planners, energy planners, city authorities and policymakers. It highlights a multidisciplinary perspective and the potential of integrated methods towards urban cooling.

Millions are moving to live in often mediocre or poor urban conditions. These new city dwellers comprise three main groups: those at the bottom of the pyramid in informal settlements, slums or low quality urban housing; a large, upwardly mobile low- to middle-income group; and a small well-off group at the top. Whilst all are important, it is in this large middle group that energy use is rising very rapidly, as they acquire amenities including air conditioning (and cars). In addition, in developing countries the buildings, vehicles and amenities these households obtain are seldom of high efficiency. Therefore, there is an ever-growing need to inform policy makers, developers, planners, designers and consultants of related sectors.

Today's rising climate emissions correspond broadly to rising energy use; this coupling will continue as long as energy supplies are largely fossil fuel based, and most projections suggest that, in 2050, two-thirds or more of global energy use will still be oil, coal and gas (US Energy Information Administration 2016). Three main solutions are available to us: conversion to renewables, energy efficiency, and downscaling or behavioural changes. Downscaling is most applicable to the rich, and is directly energy and cost saving; but people in developing countries are relatively poor and still need more, not less, basic energy amenities. Energy efficiency is often the cheapest technical option, but is not easy to achieve in developing countries due to lack of technology, resources, institutions or all three. Renewables for their part are spreading, but not fast enough to mitigate climate emissions alone. Whilst all three avenues are important, this book focuses on the task—which is preventative, and largely free—of environmentally skilful planning of cities and their overall energy systems. We do not address details of the many specific technologies, such as air conditioning or desiccant cooling. Technology is—if anything—the easier part. Sustainable design of cities can greatly reduce their needs for energy, whatever the technology; including for cooling.

Existing cities face the huge task of restructuring for sustainable development; new megacities have the opportunity to 'get it right' first time. But their rapid growth combined with a lack of resources often leads to poor solutions. Rapid growth is problematic in itself, frequently being at the expense of local environments and of living quality: 'Hong Kong's first large-scale sustainability research initiative (Barron and Steinbrecher 1999) has revealed the astonishing deterioration of the environment. The main environmental problems are associated with over-concentration due to high-rise and high-density development, and include poor air quality, water depletion, noise, and excessive waste production' (Zhang 2000). It often implies hasty, poorly prepared and controlled urban development. The field of energy is one example; even in countries, such as, Thailand and China, where adequate skills and planning systems exist, it has often been seen as too early to impose strict requirements for energy efficient buildings—which might hamper growth. To do so, in poorer regions of Asia, or Africa, is even less realistic. Hence, many cities are heading for huge future energy use and emissions, as well as, hot and unhealthy living environments.

References

Anthoff, D., Nicholls, R. J., Tol, R. S. J., & Vafeidis, A. T. (2006). *Global and regional exposure to large rises in sea-level: A sensitivity analysis.* Working paper 96. Tyndall Centre for Climate Change Research, Norwich.

Barron, W., & Steinbrecher, N. (1999). *Heading towards sustainability?* Hong Kong: Centre of Urban Planning and Environmental Management, University of Hong Kong.

Santamouris, M. (2007). Heat Island research in Europe, state of the art. *Advances in Building Energy Research, 1*, 123–150.

US Energy Information Administration. (2016, May). *World energy outlook 2016.* USDOE, Washington, DC.

USAID. (2010). *Asia Pacific regional climate change adaptation assessment: Final report findings and recommendations.* Washington, DC: USAID.

Zhang, X. Q. (2000). High-rise and high-density compact urban form: The development of Hong Kong. In J. Jenks & R. Burgess (Eds.), *Compact cities, sustainable urban form for developing countries* (pp. 244–254). London/New York: Spon Press.

CHAPTER 2

Cities, Climate and Cooling

Chris Butters and Ali Cheshmehzangi

Cities and Climate

For the first time in history, since 2007 over half the world's populations live in cities (Laski and Schellekens 2007); clustered together in communities, neighbourhoods, districts and cities that may consume as much as 75 % of global energy (Asif et al. 2007; Lehmann 2015), although occupying only 3 % of the global land surface (UNEP 2012). Buildings alone account for 40–50 % of the world's energy consumption. Developing countries have lower climate emissions per capita due to generally lower energy use, but they are set to overtake developed nations, as urban population increase will predominantly be in developing countries (Jiang and Tovey 2009). This book addresses cities, where the main energy need for indoor comfort is cooling. There are other contexts where much of our discussion is relevant. In many climates, there is some need for winter heating even if the requirement for most of the year is for cooling. Many inland continental cities, for example in central Europe and the US Midwest, have both extreme heating and extreme cooling seasons. Yet sustainable city planning and building

C. Butters (✉)
University of Warwick, Coventry, UK

A. Cheshmehzangi
The University of Nottingham Ningbo China, Ningbo, China

A. Cheshmehzangi, C. Butters (eds.), *Designing Cooler Cities*,
Palgrave Series in Asia and Pacific Studies,
https://doi.org/10.1007/978-981-10-6638-2_2

applies similar principles in all climates, if with opposite solutions—for example keeping heat out as opposed to keeping it in, or maximising cooling breezes as opposed to avoiding the chilly effect of cold winds. There will always be local or regional particularities: local weather constraints, such as, smog, rain or inversions; local opportunities, such as, prevailing breezes, water bodies and local renewable sources of energy.

There is a fundamental difference between hot-dry and hot-humid climatic conditions. Most Asian cities have hot-humid climates, whereas the Middle East and many African regions have a dry climate. In general, the hot-dry context is easier to address. In hot-dry climates, where there are often cooler nights, massive buildings help to maintain lower temperatures and this can be aided by evaporative cooling and other traditional measures; there are few such options in hot-humid climates. But in both hot-dry and hot-humid climates the two key principles for keeping cool are the same: solar protection and maximising air movement. These two apply both to the design of individual buildings and to the layout and design of streets and outdoor spaces. Whilst using these natural or 'passive' strategies can considerably reduce city temperatures, in the tropics they are often not sufficient alone and added 'active' cooling technologies are needed.

The main sources of energy use, and hence, of both heat and greenhouse gas emissions in cities are buildings and vehicles. We touch on transports only insofar as urban layout influences transport requirements. Energy for buildings—and for urban infrastructures—includes both the energy needed to operate them with cooling, lighting and so on and, importantly, the energy used to produce them. This embodied energy forms a very significant part of the overall picture; increasingly so as operational energy becomes ever more efficient. We return to this below. Space cooling in hot climates represents *the* largest energy need in buildings, often over half the total.

Energy needs are influenced by many physical factors including city density, layout, green space, building design and technology. Three main approaches to reducing them are: improved *design,* such as, passive cooling and low embodied energy materials; improving *technological efficiency*, such as, for lighting or air conditioners; and addressing energy *behaviour*. All three can be partly solved at urban rather than individual scale. Urban layout and building design can improve local microclimates, reducing ambient temperatures and hence cooling needs. District energy systems for cooling are technically much more efficient than individual ones. Planning and transport solutions as well as city management can promote energy saving and reduce travel, again reducing unwanted city heat.

Cooling in hot climates was traditionally achieved by vernacular solutions, which often demonstrate extremely intelligent 'design with nature', both for buildings and for town location and layout. Today, the universal trend is to mechanical air conditioning (AC); mostly with individual AC units in each building. Each unit ejects waste heat into the environment, thus just heating up its neighbours and increasing the energy load and urban heat island effect even more. Apart from its cost, AC can have drawbacks including noise, health risks, and the fact that it requires high quality (exergy) electrical energy. Improving the efficiency of AC is, thus, only a part solution, and helps little if, as in many cases, the *volume* of AC is increasing so rapidly that it outstrips the efficiency gains.

Good urban planning and building design provide free, passive cooling and reduce or even eliminate the need for added mechanical cooling. Passive solutions, discussed more in our examples, are the top priority. There are many reasons for this: first, to minimise the energy needed; second, they tend to be cheaper than energy supply; third, because passive solutions are robust since they are 'the building itself'; whereas all technology, renewable or not, is susceptible to lower than expected performance, incorrect use, and breakdowns. An energy efficient window is usually more cost-effective and lasts longer than a solar panel; the shading and cooling effect of trees is cheaper and longer lasting than cooling supplied by an AC unit.

Whereas technical efficiency is important, it does not solve the key issue of the urban heat island (UHI). Essentially, what is needed is to remove the sources of heat from the city environment. The key is to address cooling not at individual building level but at an urban scale instead. District energy systems, which are not yet widespread, are discussed in Chap. 11; district cooling is *almost the only* solution to UHI in hot-climate cities.

The role of green spaces and vegetation in moderating the urban microclimate is important too. They have been shown to considerably lower city temperatures, in addition to other environmental benefits, such as filtering air pollution and reducing noise. This field has been widely studied; several of our chapters provide examples of research in the Asian context. Location, layout and design of a city's green infrastructures can reduce temperatures, as well as providing outdoor wellbeing, biodiversity, leisure and social spaces.

What Kind of City?

A worldwide tradition of living in low-rise housing is giving way to life in urban apartments. This brings huge socio-cultural changes. Our paradigms of urban form include European city typologies, modernist zoning, dense high-rise, garden cities and suburban sprawl; with widely differing *economic* and *social*, as well as *ecological* characteristics. Whilst many argue for 'human scale', others praise the dynamic qualities of megacities, although few would deny that their scale and complexity imply strenuous administration and governance, including for infrastructures, transport, energy and other services. The popular 'compact city' paradigm offers advantages especially in terms of urban transport efficiency, but also a 'compact' concentration of negatives: high land prices, congestion, air pollution and noise.

Principles for sustainable building and cities are well known, but are seldom applied in a rush for development; coupled with a rather uncritical trend towards high-rise, or outdated zoning models from the modernist era. In addition comes the free rein given to private cars, with their impacts on the environmental as well as social characteristics of cities. High-rise developments are a common model, especially in the emerging economies; not only for business districts but also for residential areas. By contrast, European cities, such as London or Berlin, have pockets of high-rise but consist largely of low-rise, medium density urban fabric. At the other extreme lies the very low density 'sprawl', widespread not only in North America. It is notable, however, that high population densities can be achieved in quite low-rise cities, such as those in Europe (LSE/Eifer 2014). We have discussed elsewhere how dense high-rise entails far higher embodied energy, debatable economies of scale and few energy efficiency advantages, in addition to the many positive social qualities of low-dense type towns (Cheshmehzangi and Butters 2017). Many people believe that high-rise is a necessity in order to house growing populations, but this is not the case.

Rapid growth combined with a lack of resources often leads to poor solutions. The fast pace of development in countries, such as China, can be illustrated by the following: 'almost 80 % of the residential construction projects in Xiamen Island were built after 1990, whereas only 6.7 % of the construction was built before 1980' (Ye et al. 2011). Such rapid growth is problematic, often being at the expense of local environments and of living quality. It often implies hasty urban development. The same rapid growth applies to energy use: in Thailand, household energy use is projected to double by 2030, and over 70 % of this is in the urban sector (Chirarattananon et al. 2014); in Chinese cities like Ningbo, air conditioning (AC) in households has been increasing by 10 % per year (Ningbo Annual Statistics Yearbook 2013).

Most of the above points have particular relevance for low-income contexts, which are prevalent in many developing country cities. Whilst energy and climate emission aspects are our focus, here again we must 'join the dots' and think holistically: for example, about *economic* and *social* arguments for low-rise, including classic qualities relating to identity, security, human scale and conviviality. Dense high-rise urban environments require infrastructures and buildings of *a high standard* in order to be liveable and avoid environmental problems such as pollution, noise and UHI. In low-income contexts, which includes much of the planet, construction quality is frequently poor, which also argues against dense high-rise typologies as an appropriate model. For, whereas *high quality* compact cities may provide satisfactory conditions, *low cost* high-rise can often lead to little better than 'vertical slums'.

Over half of the world's population now lives in cities, in overall densities above 4000 persons per square kilometre; half of these live in cities of over one million inhabitants; and more than half of the largest urban areas are in Asia (Demographia 2016). We do not discuss the overall size of cities here, but regardless of size, the shaping of all cities has a large influence on energy use and climate emissions. In addition, as compared to individual buildings, which can more easily be modified or replaced, the overall city layouts and infrastructures have a very long lifetime, rendering future energy or emission improvements far more difficult. Hence, the added importance of long-term planning. City ideals have long been debated; but today's focus on energy and climate adds urgent new dimensions to planning choices. It is certainly useful to revisit known city concepts, some of which are advantageous when seen in the new light of sustainability. What layouts and spatial patterns can provide the coolest environments? The widespread paradigm of high-rise, compact cities needs to be questioned. We return to this question in the concluding section of the book.

Cooling

Heating, ventilation and air conditioning (HVAC) systems typically consume 30–50 % of the energy in domestic and commercial buildings (Gruber et al. 2008). In areas of high temperature and humidity, ventilation is one of the primary cooling tools for improving thermal comfort (Givoni 2011; Dawodu and Cheshmehzangi 2017). High energy consumption, shortages of conventional sources of energy and escalating energy prices have prompted a revaluation of conventional air-conditioning, HVAC, and design practices. There is renewed focus on passive energy design techniques to reduce energy consumption and improve thermal comfort and health whilst reducing environmental impacts (Geetha and Velraj 2012; Dili et al. 2011; Dawodu and Cheshmehzangi 2017). This view of passive design stretches beyond buildings and pertains to the entire urban environment from a neighbourhood scale (meso level) to city scale (macro level).

Factors of human wellbeing including physiological comfort are partly subjective. It is well recognised that comfort is both individual and culturally variable (Nicol 2004; Kwong et al. 2014). In the tropical regions, temperatures of around 29–30 °C are widely perceived as quite comfortable. Scientific studies build on this knowledge using tools, such as the predicted mean vote (PMV) methodology. It is also recognised that occupants feel more comfortable in spaces where they can control their indoor temperature, as opposed to automated AC environments. This can also lead to lower energy consumption. But excessive temperature and humidity cause both discomfort and ill health, especially when combined with city air pollution, and in extreme cases mortality, which is foreseen to rise in future given both rising UHI effects and global warming. Access to natural environments and green space is also recognised as important for health (Hahtela and Wolgate 2013; Hanski et al. 2012). In addition, comes reduced productivity, a major issue for workplaces including offices and not least in the many thousand factories in the Asian economies.

Both space heating and cooling involve small temperature differentials: we need to raise or lower the ambient temperatures by typically only 10–30 °C. It is, however, easier to tolerate high temperatures than it is to tolerate cold for any length of time. For thermodynamic reasons, heating is often easier technically than cooling. Available sources of free or nearly free heat in the environment are relatively plentiful; they include solar energy, waste heat from industries and geothermal (underground) heat. In cold climates, the free heat generated by lights, cooking, appliances and human bodies is also a useful source—whereas such indoor heat gains are unwelcome in hot climates. Available natural sources of cold, on the other hand, are far more limited; they can include cooler air available at night, and cool water from rivers, seas or lakes.

The physical difference between heat, work and power—thermodynamic energy quality—is central in energy planning. Power, that is electricity or motive force, is high-quality (high exergy) energy and can be simply described as highly concentrated energy; hence, more difficult to achieve. The high-quality energy available from fossil fuels in our times has enormous value. By comparison, the energy of human labour, a large component of both costs and time in developing country construction, is almost negligible when compared to most energy inputs. For example, one kilo of plastic water pipe contains around 90 MJ of energy, which corresponds to roughly 200 hours of human labour. The energy needed to construct a small timber house is around 100,000 kWh (Venkatarama Reddy and Jagadish 2003). This corresponds to around 500 years of human 'work'—a telling reminder of how incredibly useful energy is to us.

The city as a built environment is itself a cause of the heat island effect, due to its form and the hard surfaces that trap solar radiation. This is where vegetation, lakes, green roofs, reflective surfaces and other passive measures all provide cooling effects. Cities also greatly hinder the other key means of natural cooling, wind, due to the dense layout and barriers to air movement formed by the buildings. Ways to address this are discussed in several sections of the book. However, the core problem with today's predominant energy solution, mechanical AC, is that in cooling individual spaces it rejects waste heat into the city air; thus increasing the overall heat in the city. Hence, there are three generic options. The first is that of climate-adapted 'design with nature', both on the level of individual buildings and that of urban layouts; in the best case this can eliminate the need for cooling for all or most of the year. The second is technological efficiency, reducing the amount of energy needed by appliances such as AC to deliver a given comfort level. The third option is to remove unwanted heat production from the inner city altogether. This solution, district cooling, presented in Chap. 11, is at the macro level, the only way to not only stop increases in city heat but to reduce it. It can be added that the same applies to the other major source of city heat, namely vehicles; to reduce this, one must remove the source of the heat from the inner city, either by reducing the volume of traffic or by replacing it as far as possible with electric and other non-heat producing transport modes.

Both passive design and natural ventilation are fast developing fields today; as is that of district cooling. In short, solutions exist but they are not widespread. The urgency of this is well recognised in Asia including within major organisations such as ASEAN; again, the main challenges are not technical but ones of resources, policy and implementation.

Energy and Carbon

For readers less familiar with this field, we include brief notes here on energy versus carbon and operational versus embodied energy (OE, EE). In this book we refer generally to energy; the *carbon* (or climate) impacts of built environment correspond broadly to those of *energy* since energy supply systems are largely fossil-fuel based. A reduction of the *carbon* impact with significant *decoupling* occurs only when our energy supplies become largely renewable.

Energy is both a cornerstone of our lives and important due to its costs and impacts. Energy reductions generate cost savings, for people, and greenhouse gas savings, of global benefit. Energy is used in creating buildings and cities and again in operating them. The *carbon* angle is important due to climate change; carbon saved now is more 'valuable' than carbon saved in 100 years' time, by which point one assumes the climate issue will have been solved (or else it is too late anyway!). Energy is a permanent challenge, but even more today due to its correlation with the climate question. The energy footprint of buildings and other products should diminish over time as production processes become more efficient and as our energy is more and more from renewable sources. In the context of buildings and city infrastructures a major exception to this is the case of Portland cements, where the *chemical process* of production (calcination) emits more CO_2 than the production energy, and is a large factor in the lifetime footprint of many constructions. This one material is the source of roughly 5 % of global carbon emissions (Worrell et al. 2001). For sustainable building, we must in addition, consider toxicity to humans or the environment; this includes, substances that may not be important in terms of energy or climate (including SO_2, NOx, VOCs, formaldehyde, synthetic mineral fibres, lead, asbestos and the like).

Until recently, in colder climates, such as Europe, the energy used to heat buildings was by far the largest item. With recent low energy buildings, heat loss has been reduced to a fraction; electrical appliances, including, lighting are also now far less energy consuming. Since the *operational* energy is thus greatly reduced, the *embodied* energy to produce the materials becomes a far larger part of the total. This picture has only emerged over the last 15–20 years. The part played by the *materials* assumes great importance.

The fields of OE and EE are well established but complex. Methodologies, assumptions and contexts require careful handling. The prevalent approach is life cycle analysis (LCA), which also has variants; adopted boundaries may be cradle-to-gate, cradle-to-grave or cradle-to-cradle—as well as limitations; for example the post-use phase does not account satisfactorily for residual energy/carbon (Sassi 2008). The focus of such studies also depends on the purpose; the perspective (and goals) of a manufacturer is different from that of a product designer, and different again from that of a climate scientist.

For buildings, and to a large extent for other products too, OE is most commonly accounted in terms of MJ or kWh, measured per area of building, to which one may add per year of lifetime or per whole lifetime. It is notable that building lifetime is often assumed to be 50 years, both in legal standards and for cost-benefit calculations. This, in our opinion, is too short to be called 'sustainable'. The lifetime assumption obviously has a big impact on the result. EE is similarly accounted in terms of MJ per kg of building material, eventually aggregated as MJ per unit area of building. For carbon, corresponding units are normally kg CO_2e per unit area or per kg of product. All other greenhouse gases (GHG) being converted to CO_2 equivalents; for example, 1 kg of methane has the same greenhouse effect as about 21 kg of CO_2. One thus arrives at an overall figure for a building's performance, which can be compared to other buildings, used to set up benchmarks or, in the design stage, to evaluate initial concepts and develop more favourable options.

Embodied energy and carbon figures vary widely. For example, aluminium made using renewable hydropower in Norway requires the same production *energy* per kg but has a much lower embodied *carbon* impact than aluminium that is produced in China using fossil fuels. If we use a database containing figures based on 'EU average' for aluminium production, we cannot directly compare to a building where the aluminium is produced using energy from oil or coal. Similarly, in developing countries, factory efficiencies may be much lower than those in, say, Europe. Hence, one must often revert to a detailed consideration of the *primary energy* picture.

Operational energy is the energy needed to run buildings. OE includes space climatisation, cooking, hot water, lighting and other appliances. In low-income contexts, it is very important because it requires a cash flow. In some developing countries cooking forms the largest home energy demand. In the rapidly urbanising low- to middle-income sectors, space cooling forms the main growth in energy use. As noted it can to a large extent be ensured by good building design using passive means, less easily in hot-humid climates, and not in polluted cities where outdoor air is undesirable. Being free, passive solutions are a priority.

Embodied Energy (or carbon) has received less focus. In low-energy buildings, the EE may constitute over half the *total* lifetime energy. A building's life has three main phases: construction, recurrent maintenance or modification, and final demolition. EE is the total energy used over the building lifetime for manufacturing its components, transport, assembly on site, lifetime maintenance and final post-use disposal. Production of the materials normally accounts for the bulk of the EE. Vernacular building uses natural materials like earth, stone, timber and other vegetal products. These have environmental advantages, such as low energy intensity, local availability and biodegradability. With energy and climate in focus, the traditional materials are regaining interest.

Means of reducing EE include selecting less energy-intensive materials, using less material, increasing the efficiency of material manufacture and extending building life. In hot-dry climates, heavy materials are often favourable, but lightweight solutions in hot-humid climates. Lightweight means generally less material and lower EE: an inherent carbon advantage. But urban buildings today are increasingly of heavy materials, such as concrete and steel. Looking ahead, the challenge of reducing EE is equally relevant for less developed contexts, since urban construction is following similar energy—and carbon-intensive trends all over the world. One can fairly easily reduce EE by up to 50 % in both hot-dry and tropical climates through use of lower EE materials. For many though not all purposes, revitalisation of traditional low-energy materials, in improved forms, is advisable, including non-cement-based masonry and new plant-based biomaterials.

Integrated Solutions

Since mitigation of energy use and climate emissions is only partly about individual buildings but equally about the overall city planning and energy supply systems, there is a great need to link engineering and architecture with landscape design, urban planning and energy planning. Sustainable solutions demand whole thinking and integrated planning processes. To simplify, one may say that the architect is looking at the individual building, the urban planner only at the overall layout and the energy planner only at issues of energy supply. In energy planning, demand side management

(DSM) receives increased attention today, but there is still little consideration as to which of our three levels offers the best solutions, or in what combination. Not only the skills but the focus and interests of the actors in these fields are often divergent or even conflicting. We refer to the issue of integration at several points in the book.

There are many synergy effects to be achieved by seeing building, urban layout and city energy system in conjunction. And we should not confine our view to the city limits. As has been illustrated in an Ecocity Master Plan developed by GAIA architects for another Asian context, Taiwan, there are synergy effects to be realised by integrating a city and its hinterland (Butters 2013; Bokalders and Block 2007). In regard to city cooling, it is mainly environments outside the city itself—air, river or others—that are the recipient or heat sink for the large quantities of waste heat that arise when producing energy including district cooling to the city. The surrounding countryside, our source of food, energy, water, wastes treatment and recreation—and cool fresh air—must be understood and revitalised as integral part of a total eco-social system. This must be the paradigm for sustainable human settlements. There is really no such thing as a 'sustainable city', a term which we argue is in some ways an oxymoron (Butters 2013; Cheshmehzangi 2016).

This is another reason why we have combined design practice, research, planning and policy in this book. The specialist approach often leads to missed opportunities. The relatively new area of 'Industrial Ecology' is an example of holistic planning, where industries and other functions are co-located and planned so that both the by-products and waste energy streams from each part are used as resources in others. As will be shown in Chap. 11, some district energy systems in countries like Malaysia are examples of this integration.

Some Underlying Issues

Sustainable city development demands broad vision; the topic of urban cooling ties in with several quite complex and interconnected questions.

Energy and Poverty

As noted, it is the low-income groups who risk being especially disadvantaged in the hot megacities. There is some conflict between two equally valid policy goals: *reductions of energy use and climate emissions* on the one hand, and *poverty alleviation* on the other. Absolute reductions in energy use or climate emissions cannot be demanded of the poor. The rich can reduce their footprint, by efficiency gains and reduced consumption; but the poor need access to *more energy*—more cooling as well as more space, lighting, transport, and public services. The greatest growth in energy use and climate emissions is in the low- to middle-income segment of the upwardly mobile urbanising populations. This tendency is worldwide and likely to persist; hence, our topic is of growing relevance in the near future for less developed countries too. Must the new urbanising millions follow a conventional development path, in unhealthy conditions in conventionally designed cities, towards the excessive energy and resource use of the rich? Can we reduce this coming growth in energy use? Our energy/climate goal must be largely preventative; not *absolute reductions*, but *avoided future impacts* of these cities.

Context, Technology and Behaviour

Second, energy including space cooling is relative to socio-economic factors. Whilst the rich in all countries tend to have similar energy amenities, low-income groups may still have no energy amenities at all. Hence, our focus will depend on the particular city context, both climatic, socio-economic and cultural. In terms of needs and equity, the poor are most important. Priorities for energy policy, planning and buildings depend on specific socio-economic conditions in the urban districts in question. Similarly, there is wide recognition today that energy is not just a technical matter, but a socio-technical one. Energy behaviour—such as the much-discussed rebound effect, and even 'prebound effect' (Sunikka-Blank and Galvin 2012)—is increasingly in focus; many energy-efficiency programmes are not achieving the calculated results due to behavioural factors. Poorly chosen solutions may result in high energy use due to such factors despite being technically very efficient. This signals the need for a major shift in energy-policy approaches, to a better balance between the technical and sociological aspects; again a question of cross-disciplinary understanding and integration.

Defining Comfort

Third, indoor environment and thermal comfort are defined in differing ways. It is well known that 'advanced' norms and standards, such as those of the World Health Organisation (WHO) or the American Society of Heating, Refrigerating and Air-Conditioning Engineers (ASHRAE), most of which were developed in the West, are not always applicable in hot climate cultures—where higher temperatures and humidity may be perceived as comfortable (Wang et al. 2010). Those are 'high end' norms, which imply high energy use and will, in addition, be unrealistically costly to achieve in many developing country contexts. Given the lack of resources in much of the Asian context, if we insist on absolute norms then the poor are likely to stay as hot and uncomfortable as they are for many years to come. In the poorer cities or typical slum contexts, which have neither energy access nor money, our primary goal should rather be a pragmatic one: to *improve* indoor (as well as outdoor) thermal conditions significantly, but without added costs; not necessarily to achieve the unrealistic goal, at least in the short term, of *absolute* standards or absolute reductions in climate emissions.

Research and the Real World

Finally, there exists a large body of scientific and research literature on cities and urban energy. However, many of the specialist scientific publications are not open access and are seldom read outside academic circles. Knowledge on passive cooling has been around for several decades but is seldom being applied in contemporary city planning. This underlines a frequent disconnect between research and reality. Much research is very theoretical too, whereas there are quite simple principles and solutions available, requiring little more than awareness and knowledge, which can be applied today by city authorities and planners. Hence the need for a pragmatic approach; ideal solutions are seldom attainable in the real world of cities. The need is not so much for technical innovation as for dissemination and delivery of tried and known solutions.

We now turn to nine selected case studies across the hot climate zones of Asia, three on each level of micro, meso and macro considerations towards cooler cities.

References

Asif, M., Muneer, T., & Kelley, R. (2007). Life cycle assessment: A case study of a dwelling home in Scotland. *Building and Environment, 42*, 1391–1394.

Bokalders, V., & Block, M. (2007). GAIA international: Tainan Ecocity. Sustainable municipal and building, *World Architecture* 07/2007, 205, Japan.

Butters, C. (2013). Is there such a thing as an Ecocity? From Ecocity to sustainable habitat, *Scottish ecological design association magazine SEDA*, Edinburgh, Autumn 2013.

Cheshmehzangi, A. (2016). City enhancement beyond the notion of 'sustainable city': Introduction to Integrated Assessment for City Enhancement (iACE) toolkit. *Energy Procedia, 104*, 153–158.

Cheshmehzangi, A., & Butters, C. (2017). Chinese urban residential blocks: Towards improved environmental and living qualities. *Journal of Urban Design International, 22*(3), 219–235.

Chirarattananon, S., Rakkwamsuk, P., Chiarakorn, S., Chaiwiwatworakul, P., & Chuangchote, S. (2014). Review of features of and energy use in medium and low income housings in Thailand, Joint Graduate School of Energy and Environment (JGSEE), King Mongkut's University of Technology Thonburi (KMUTT). Interim report, ELITH program.

Dawodu, A., & Cheshmehzangi, A. (2017). Passive cooling energy systems SWOT analyses for energy-use reductions at three spatial levels. *Energy Procedia, 105*, 3411–3418.

Demographia. (2016). *World urban areas. 12th annual edition*: 2016:04, Available at: https://www.demographia.com/db-worldua-index.htm. Accessed 5 May 2017.

Dili, A. S., Naseer, M. A., & Zacharia Varghese, T. (2011). Passive control methods for a comfortable indoor environment: Comparative investigation of traditional and modern architecture of Kerala in summer. *Energy and Buildings, 43*, 653–664.

Geetha, N. B., & Velraj, R. (2012). Passive cooling methods for energy efficient buildings with and without thermal energy storage – A review. *Energy Education Science and Technology Part A: Energy Science and Research, 29*(2), 913–946.

Givoni, B. (2011). Indoor temperature reduction by passive cooling systems. *Solar Energy, 85*(8), 1692–1726.

Gruber, E., Plesser, S., Dusée, R., Sofronis, I., Lima, P., & Rivière, P. (2008). Deliverable D26: Report on the project results. Technical report, EL-TERTIARY project, intelligent energy – Europe, August 2008.

Haahtela, T., Holgate, S., Pawankar, R., Akdis, C. A., Benjaponpitak, S., Caraballo, L., Demain, J., Portnoy, J., & von Hertzen, L. (2013). The biodiversity hypothesis and allergic disease: World allergy organization position statement. *World Allergy Organization Journal, 6*(1), 3. PMCID: PMC3646540.

Hanski, I., von Hertzen, L., Fyhrquist, N., Koskinen, K., Torppa, K., Laatikainen, T., Karisola, P., Auvinen, P., Paulin, L., Makela, M. J., Vartiainen, E., Kosunen, T. U., Alenius, H., & Haahtela, T. (2012). Environmental biodiversity, human microbiota, and allergy are interrelated. *Proceedings of the National Academy of Sciences, USA, 109*(21), 8334–8339.

Jiang, P., & Keith Tovey, N. (2009). Opportunities for low carbon sustainability in large commercial buildings in China. *Energy Policy, 37*(11), 4949–4958.

Kwong, Q. J., Adam, N. M., & Sahari, B. B. (2014). Thermal comfort assessment and potential for energy efficiency enhancement in modern tropical buildings: A review. *Energy and Buildings, 68*, 547–557.

Laski, L., & Schellekens, S. (2007). Growing up urban. In A. Marshaland & A. Singer (Eds.), *The state of world population 2007 youth supplement*. New York: United Nations Population Fund (UNFPA).

Lehmann, S. (2015). *Low carbon cities: Transforming urban systems*. London/New York: Routledge.

LSE Cities/Eifer. (2014). *Cities and energy urban morphology and heat energy demand*. Final project report, London School of Economics, London/Karlsruhe.

Nicol, F. (2004). Adaptive thermal comfort standards in the hot–humid tropics. *Energy and Buildings, 36*, 628–637.

Ningbo Annual Statistics Yearbook. (2013). China statistics press.

Sassi, P. (2008). Defining closed-loop material cycle construction. *Building Research & Information, 36*(5), 509–519. UK.

Sunikka-Blank, M., & Galvin, R. (2012). Introducing the prebound effect: The gap between performance and actual energy consumption. *Building Research & Information, 40*(3), 260–273, UK. Accessible from: https://doi.org/10.1080/09613218.2012.690952. Accessed 2 June 2017.

UNEP. (2012). *Global initiative for resource efficient cities*. UNEP: Paris.

Venkatarama Reddy, B. V., & Jagadish, K. S. (2003). Embodied energy of common and alternative building materials and technologies. *Energy and Buildings, 35*, 129–137.

Wang, Z., Zhang, L., Zhao, J., & He, Y. (2010). Thermal comfort for naturally ventilated residential buildings in Harbin. *Energy and Buildings, 42*, 2406–2415.

Worrell, E., Price, L., Martin, N., Hendriks, C., & Meida, L. O. (2001). Carbon dioxide emissions from the global cement industry. *Annual Review of Energy and Environment, 26*, 303–329.

Ye, H., Wang, K., Zhao, X., Chen, F., Li, X., & Pan, L. (2011). Relationship between construction characteristics and carbon emissions from urban household operational energy usage. *Energy and Buildings, 43*, 147–152.

PART I

Case Study Chapters: Micro Level

In this section of the book we present three cases that address the micro level of individual buildings. The first is an example of ongoing applied research in a state-of-the-art building in China. This is followed by a design study of a climate-responsive building in tropical Sri Lanka, where we also highlight vernacular or traditional approaches to creating cooler indoor environments. The third chapter in this Part addresses the topic of national policy and regulations towards energy-efficient and cooler buildings, taking the case of a project led by the editors' research colleagues in Thailand.

CHAPTER 3

Reducing Cooling Loads in Hot-Humid Climates: A Best Practice Research Building in China

Ali Cheshmehzangi and Linjun Xie

Introduction

In this chapter we discuss a project that epitomises the research required to find sustainable solutions and create comfortable environments at the micro scale of individual buildings, the IBR centre in Shenzhen, China, which was a research project in itself and continues to be a focus of ongoing research, experimentation and monitoring. The knowledge base ranges from the sciences of climatology, human comfort and behaviour to those of building physics, energy systems and advanced engineering; the challenge lies in integrating these diverse fields. Whether at micro, meso or macro level, the common cause is the application of advanced design and science to solutions in practice of our overarching challenge: warming environments, both at the local level of cities and that of the planet.

Since the Kyoto protocol of 1997, most countries have committed to reducing fossil fuel consumption and greenhouse gas emissions. Buildings outweigh both industry and transport and account for some 40 % of the total picture (WBCSD 2007), with further increases due to rising world

A. Cheshmehzangi (✉) • L. Xie
The University of Nottingham Ningbo China, Ningbo, China

A. Cheshmehzangi, C. Butters (eds.), *Designing Cooler Cities*,
Palgrave Series in Asia and Pacific Studies,
https://doi.org/10.1007/978-981-10-6638-2_3

population and standards of living (Pérez-Lombard et al. 2008; Kamal 2012). With global emissions from buildings predicted to reach 42.4 billion metric tons by 2035, 43 % more than in 2007 (USEIA 2010), energy saving and efficiency in buildings has become a primary policy (Yilmaz 2007). Many variants of green or sustainable building illustrate the advances made in design, technical and managerial terms. With a variety of assessment tools now adopted to guide green building design and construction (Leadership in Energy and Environmental Design (LEED) in the USA, Building Research Establishment Environmental Assessment Method (BREEAM) in the UK, Green Building Council of Australia (GBCA) in Australia …), energy efficiency is one of the key criteria (Zuo and Zhao 2014).

The HVAC (heating, ventilation and air conditioning) systems in buildings are the largest energy user amongst all energy services (Pérez-Lombard et al. 2008). Heating and cooling loads are the 'measure of energy needed to be added or removed from a space by the HVAC system to provide the desired level of comfort' (Burdick 2011, p. 1). In the hot-humid climate zone, cooling needs dominate. Main factors contributing to the cooling loads are heat gains through opaque and glazed external surfaces and internal heat gains (Aktacir et al. 2010; Wang 2000). External sources are principally solar gain and infiltration of hot outdoor air; internal sources are principally from people, cooking, lighting, and electrical appliances (ibid.). The architectural and physical properties of buildings such as their design, thermal mass and materials largely determine the cooling needs (Aktacir et al. 2010). With buildings being both the thermal enclosure towards the outdoors and a source of internal heat, measures to reduce cooling load are closely linked to the improvement of building design and components (Burdick 2011).

Reduction of Cooling Loads in Hot-Humid Climates

Techniques to reduce the cooling load are well known. Protection from the sun and maximum air movement are the two most basic 'defensive' strategies; in addition to minimising internal heat production. The appropriate use of thermal mass—materials that can absorb, store and release heat—can reduce the cooling load (Yang and Li 2008; Balaras 1996) especially when combined with night ventilation to cool the building fabric (Kolokotroni and Aronis 1999; Yang and Li 2008). Design-enhanced ventilation is a primary and low-cost technique to provide thermal comfort, in addition to a healthy indoor environment (Blondeau et al. 1997). Hui (2007) states that passive cooling by induced natural ventilation is a most effective approach in hot-humid climates. At the same time, it is well known that human thermal comfort varies in different climate conditions and cultures (Givoni 1998). Hyde (2000) notes that the basic philosophy of climate-responsive design lies in optimisation of building performance in relation to local climate; hence the need for a full understanding of climatic influences.

Hot-humid climates pose unique challenges due to the combination of internal heat, intense solar conditions and high relative humidity, rendering conventional methods of reducing energy use difficult (Parker et al. 1997). For example, extremely humid conditions can affect the insulating layer and increase the rate of heat transfer through the building envelope, increasing the cooling loads (Wang 2000). In an analysis of thermal comfort in residential buildings in Malaysia, Zain et al. (2007) highlight the characteristics of the solar radiation and the building envelope as well as ways to improve the microclimate in the surrounds of buildings. Bojic et al. (2001) found that even in hot-humid regions, thermal insulation on the outside walls can reduce the annual cooling needs as well as peak cooling demand. A similar conclusion is found by Aktacir et al. (2010) in a case study on the performance of building thermal insulation in Turkey.

Characteristics of Hot-Humid Climate Zones in China

China's hot-humid climate zones occupy one-third of the country including the southwest, the middle and lower reaches of the Yangtze River and the south, with the Sichuan Basin, part of the Guizhou region, and 21 provinces and regions (such as Guangdong, Chongqing, Fujian, Hunan, Hubei, Jiangsu, Zhejiang, and Anhui provinces) (Fig. 3.1) (Deng and Li 2016). This climate features high annual rainfall, humidity and temperature, strong sunlight, and violent lightning (Xie and Liu 2006; Sun et al. 2014). Average temperature is around about 27 °C whilst relative humidity is 80–90 % for most of the year (Sun et al. 2014). Research studies show that traditional buildings in these parts of China exhibit distinct features and effective design responding to the climatic conditions (Xie and Liu 2006; Wang 2011). Deng and Li (2016) summarise the natural ventilation strategies adopted by these traditional dwellings. For instance, the buildings are usually small-scale, well shaded, with a large roof to facilitate ventilation and cooling, and using materials with good permeability maximising cross-ventilation. These strategies continue to inspire architects and engineers and are applied in modern building design (Sun et al. 2014). Research on buildings in the Chinese hot-humid climatic zone includes general design

Fig. 3.1 IBR headquarters building in Shenzhen, south façade (Source: The Authors)

principles (Fang and Cheng 2016; Sun et al. 2014), detail design such as the optimisation of residential peripheral structures (Liu et al. 2015), and HVAC design (Liang and Huang 2005).

The Shenzhen IBR Building in South China

The Shenzhen Institute of Building Research (IBR) is a well-known research institute in the city of Shenzhen, Guangdong Province. Since its establishment in 1982, IBR has completed many local projects and cutting-edge research with international collaborators. In recent years their focus has been on the 'green life concept in China' addressing green development and lifestyle. Their new headquarter building is itself a 'green experiment' (Malone 2010). Located in the Futian District of Shenzhen, it is often highlighted as a model for China's future. Construction was completed in 2009. Occupied by over 400 persons, main activities are design and research including several floors of offices and research labs, open spaces at various floors, servers for energy data, an auditorium, meeting spaces, an underground car park, a nursery, recreation and lifestyle facilities. With a total cost of $12.9 Million USD, the building has 12 floors above ground and 2 floors of basement functioning as part of the building's R&D platform. The gross floor area is approximately 18,000 m^2. The building has an overall vertical layout with optimised air circulation and environmental quality (Fig. 3.1). The design includes more than 40 sustainable technologies and many passive cooling strategies.

A relatively low-cost and very low-energy building, IBR HQ has achieved many distinguished honours and awards, which underline its achievements and performance as evaluated from many points of view. (see later).

In a climate where cooling is normally needed for over six months per year, the building's integrated green features and innovative sustainable technologies decrease emissions by some 767 tons of CO_2 annually, corresponding to over 400 tce (tons coal equivalent) (Cao et al. 2014; Diamond et al., ASHRAE Report 2014). An assessment conducted by ASHRAE (2014, in Diamond et al. 2014) indicates an annual energy use of 63 kWh/m^2, which is little more than half the average of other recent office buildings in the region. This figure, incidentally, is comparable to that of 'passivhaus' type low-energy buildings in the cold climates of Europe. Based on post-occupancy evaluation (Diamond et al. 2014), these results are mainly due to the building's use of cooling strategies such as natural ventilation and integrated daylighting design. We now discuss some of the key design principles integrated into the building that lead to its exceptional cooling performance.

Sustainable Features and Technologies in IBR Building

The development of the design for IBR is strongly based on local climatic analyses, focusing on natural ventilation and daylighting principles specifically for Shenzhen's hot summer and warm winter climate. The team then conducted computational modelling to optimise the location, size, shape and building envelope; considering where and how to optimise direct airflow through openings and interior spaces. The most prominent aspects of the integrated design strategy are as follows:

Layout and Orientation The overall building design is largely based on computational fluid dynamics (CFD) simulations of Shenzhen's main prevailing wind directions, south-east to north-west, with emphasis on cross-ventilation and maximisation of airflow in all floor spaces. The most visible design feature is the intentionally asymmetrical façade which both steers and enhances air flows into the building's main core and open spaces; especially in function of the South-East breezes that are predominant during the hottest season. The focus on air flows also provided an opportunity for attractive and functional open spaces on all floors. Three key design features are: a large central open space dividing the building layout into two wings; a setback of the south wing, which directs more airflow into the central space; and a thinner south wing specifically designed to deviate more of the airflow coming from the south-east in the hot season. Figure 3.2 demonstrates how these features are integrated in the building layout.

Building Envelope In the vicinity, a rapidly growing business quarter is reshaping the existing urban fabric, with many mid- to high-rise office buildings in the area. IBR has significant differences. Unlike most it does not have a fully glazed façade on the lower floors but has operable windows and balconies, and a setback to reduce direct sunlight into the indoor spaces. Whereas most buildings are the same on all sides, the design team focused on detailing of windows and curtain walls so that each façade has a different strategy for the optimisation of fixed and operable windows (Diamond and Feng 2014). For instance, the building's western façade includes a layer of solar photovoltaic (PV) panels, which reduces afternoon direct sunlight and provides renewable energy production (ibid., p. 48). Similarly,

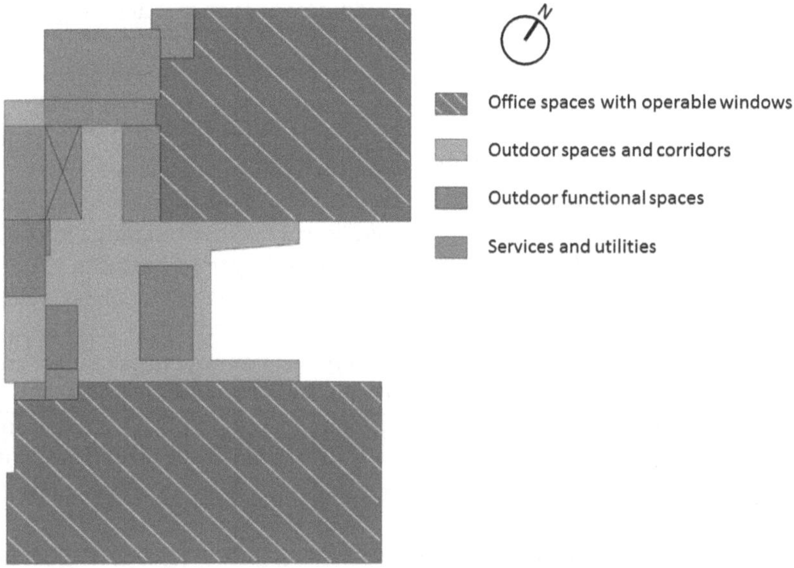

Fig. 3.2 Schematic diagram of building layout at upper floors of IBR HQ building (The Authors, adapted from Cao et al. 2014)

the rooftop is covered by PVs and provides shading for the outdoor atrium. Through computational modeling (CFD), climate strategies where simulated for each façade, with shading devices and overheating reduction measures. Moreover, IBR HQ has a different thermal envelope on each façade. To maintain enough shading, different window-to-wall ratios (WWR) are applied for different areas and façades. For instance, the lower floors have a WWR value of 0.3 for the south, east and north façades whilst the upper floors have a WWR value of 0.7 (ibid., pp. 23–24). The former reduces daylight impact on meeting rooms, auditorium and lab spaces, whilst the latter maximises natural daylight for office spaces. Shading devices are also integrated in the exterior wall finishing, which again differs from lower to upper floors. With high performative thermal integrity, the building façades are constructed with aluminium cladding and insulation materials, creating an energy-efficient building envelope.

Spatial Use The strategy behind the building's indoor spatial arrangement adds significantly to its cooling performance. The combination of a raft foundation (unlike the common practice of pile foundations in Shenzhen) and underground spaces enhances the overall airflow from the building's basement spaces. A major climatic feature of the building is the greening and plantation strategy at multiple levels, highlighting native species, vertical plantation on certain areas of façades, the planted roof space (sky garden), and a planted space in the middle part of the building. Another major feature of the building is its open plan with open walkways at various levels, functional open spaces for meetings, eating and interaction, as well as an open lobby area facing the summer wind direction. These air flow and layout strategies minimise indoor pollution and risks of condensation and fungal growth (Diamond and Feng 2014, p. 22). The main features such as the sky garden (Fig. 3.3 top), planted balcony spaces (Fig. 3.3 bottom), and flexible indoor spaces with moveable walls, create mostly daylit and naturally ventilated workspace environments.

Ventilation In addition to its integrated design strategies, IBR benefits from energy efficient features, including its natural ventilation, HVAC systems, and sustainable technologies; The horizontally pivoted windows maximise direct airflow into the building; the central atrium constitutes a large open 'breathing space' between north and south wings, whilst most of the less favourable western side of the building is utilised for services, open corridors and circulation. This layout also maximises cross-ventilation from east to west, vital for the cooling performance. Both the open lobby area and the atrium provide optimal natural ventilation to offices, reducing the use of air conditioning. As a result, most service areas, toilets, elevators and balcony spaces rely exclusively on natural ventilation (Diamond et al. 2013). Unlike many office buildings in the region, mechanical cooling is kept to a minimum and is only partially used when outside air temperature reaches above 25 °C. Based on a comfort-assessment study, conducted by Diamond et al. (2013), the level of occupants' satisfaction (identified as 'comfortable') was 88 % for thermal comfort and 75 % for indoor humidity, in the naturally ventilated mode.

Besides the natural ventilation strategies, the building's HVAC system, located in the basement and first floor is also different from other offices in the region. The lower floors use a water source heat pump (WSHP) for summer cooling when energy use reaches its peak, doubling from 3000 kWh/day in winter to 6000 kWh/day in summer (Diamond et al.

Fig. 3.3 Top: Green spaces i of the sky garden. Bottom: planted balcony spaces on upper floors (Source: The Authors)

2013). The closed loop condenser of the heat pump is positioned next to the landscape water pool outside the building and exchanges heat directly with the pool. The upper floors use water chillers with solution-based dehumidification air handling units (ibid.). Mechanical cooling is in partial operation for around half of the year.

Discussion: Design and Practice

In practice, IBR can be regarded as a living green laboratory; a building with advanced sustainable technologies, contemporary design approach and adequate energy-efficient techniques. Such integrated design benefits greatly from the CFD modelling studies, to optimise overall building layout, functional zoning, structural design, façade design and window shapes and locations. This enables the designers to perform comparative studies. Not only that CFD is 'an effective tool for simulation and calculation of part of the environmental performance in the built environment' (Cheshmehzangi 2016, p. 1086), but it also provides a set of valuable information at the pre-assessment phase and can be utilised as an analytical tool (ibid.).

The figure cited above of annual energy use, around 63 kWh/m^2, offers an interesting perspective. This figure is very comparable to the performance of state-of-the-art low-energy buildings in the temperate and cold climates, such as the 'passive house' type that is fast spreading in Europe. In essence, these advances are very similar. The component of energy use for space climatisation (either cooling or heating) used to be the major part of total energy needs; in both cases it is this part that is being dramatically reduced. The other main parts include the energy needed for lighting and electrical appliances; these are becoming far more efficient, such as LED lighting, although the number of appliances is increasing. The internal heat produced by the occupants is now also a significant factor in low-energy buildings. Whereas the heating need in colder climates has been dramatically reduced by thicker insulation, insulating windows, heat recovery systems and other features, the cooling load in hot climates is dramatically reduced by better passive design as well as more efficient cooling technologies—as in the IBR building. Further, the low remaining need for energy is then covered as far as possible by renewable energy production on site, such as PV solar electricity, in the case of buildings in both hot and cold climates. One therefore arrives at 'net zero energy' buildings. In both climates, the embodied energy (or carbon) of the construction materials is also receiving much attention. Indoor comfort and environment is in both cases the core objective; whether keeping warm or keeping cool.

Providing cooler building design (not only aesthetically) through better performance and energy efficiency, may in some cases appear costly but should be seen from the perspective of the long-term benefits, such as energy use reduction, quality design, better thermal comfort, and not least better health and higher productivity. IBR achieves these in practice. As a pioneer in the context of south China, IBR's cooling performance indicates the success of the design methods, technology and application of these in practice.

Conclusions

Between 2007 and 2011, the prominent nationwide programme '100 Demonstration Projects of Green Buildings and 100 Demonstration Projects of Low-energy Consumption Buildings' was conducted by the Chinese Ministry of Housing and Urban-Rural Development (MOHURD). Since then, MOHURD has taken a lead in green building evaluation and certification, applying both national and international labelling programmes such as Green Star, BREEAM, LEED, CASBEE, etc. Although new for China, in this period a total of 271 buildings were awarded with a green building label (Shui and Li 2012). IBR was one of the first that was evaluated and was recognised as one of the most successful early models. Since then, green building practice has been spreading, and has become or soon will be compulsory in many Chinese cities. In Shenzhen it is already compulsory for new public buildings; a move that has changed current practice for new building projects.

This building is in many ways a live research laboratory. Even as the building is in use, different facades are being tested in various modes on an ongoing basis, as are control and indoor environment management tools. New approaches have been greatly facilitated by the development of advanced simulation techniques. In this way, future solutions to global broad policy goals are being developed and are hopefully to soon become widely applied building practice. In specific terms, the case of the IBR building highlights the effectiveness of integrated methods of cooling, especially passive design strategies with natural ventilation and innovative design.

Acknowledgment We are particularly grateful to the project team at Shenzhen Institute of Building Research (IBR) for giving us access to the building and to some useful documents on the building's performance and design.

Notes

The IBR building has achieved the following awards and honours:

First Grade Distinction (highest score) First place in the National top 100 Green Building demonstration Projects (China 2010), 2010 Hong Kong Building Award (Merit Level), 2011 China Human Settlement Pattern Project Award (2011), highest National Green Building Award (3 Star, China 2010), First place in the National Demonstration Project of Renewable Energy Application (China 2010), 2011 National Excellent Engineering Design Award (First prize, China), FutureArc Green Leadership Award 2011 (by BCI Asia), 2013 World Green Design Project Product Award (Golden Prize by International Design Federation), DNA Energy Efficiency Award 2013 Nomination, Asia-Pacific Award by World Green Building Council (WGBC), as well as, the third Biannual top Architecture in the Public Building Category, Green and Ecological Design Award, the third Biannual China Award for 'Good Design is Good Business', and Best Green Design Award (by Business Week and McGraw-Hill Construction) (Diamond et al., ASHRAE Report 2014).

References

Aktacir, M. A., Büyükalaca, O., & Yilmaz, T. (2010). A case study for influence of building thermal insulation on cooling load and air-conditioning system in the hot and humid regions. *Applied Energy, 87*, 599–607.

Balaras, C. A. (1996). The role of thermal mass on the cooling load of buildings. An overview of computational methods. *Energy and Buildings, 24*, 1–10.

Blondeau, P., Spérandio, M., & Allard, F. (1997). Night ventilation for building cooling in summer. *Solar Energy, 61*, 327–335.

Bojic, M., Yik, F., & Sat, P. (2001). Influence of thermal insulation position in building envelope on the space cooling of high-rise residential buildings in Hong Kong. *Energy and Buildings, 33*, 569–581.

Burdick, A. (2011). *Strategy guideline: Accurate heating and cooling load calculations.* Washington, DC: IBACOS, US Department of Energy.

Cao, B., Li, M., Liu, G. & Zhu, Y. (2014). *Thermal comfort in an open space of an office building: A field study in subtropical region.* In proceedings of the 13th international conference on indoor air quality and climate, at Hong Kong, topic A7: Thermal comfort.

Cheshmehzangi, A. (2016). Multi-spatial environmental performance evaluation towards integrated urban design: A procedural approach with computational simulations. *Journal of Cleaner Production, 139*, 1085–1093.

Deng, S., & Li, J. (2016). Natural ventilation strategy of traditional dwellings in hot-humid area. *Architecture & Culture, 8*, 108–110.

Diamond, R. C. & Feng, W. (2014). *Sustainability by design: How one building in China could change the world.* Published by the regents of the University of California, through the Lawrence Berkeley National Laboratory.

Diamond, R. C., Ye, Q., Feng, W., Yan, T., Mao, H., & Li, Y. (2013). Sustainable building in China – A green leap forward? *Buildings, 3*, 639–658.

Diamond, R. C., Feng, W., & Qing, Y. (2014). Model for China's future, ASHRAE report document. *High Performing Buildings*, 7(3), Summer 2014, ASHRAE Publications.

Fang, L., & Cheng, Y. (2016). Passive design for heat proof and energy efficiency of building in hot and humid area: Analysis on buildings in Guangzhou. *Building Energy Efficiency, 10*(44), 65–67.

Givoni, B. (1998). *Climate considerations in building and urban design.* New York: Van Nostrand Reinhold.

Hui, S. C. M. (2007). *Sustainable building technologies for hot and humid climates.* Invited paper for the Joint Hong Kong and Hangzhou seminar for sustainable building. Hangzhou, China.

Hyde, R. (2000). *Climate responsive design: A study of buildings in moderate and hot humid climates.* London: E & FN Spon.

Kamal, M. A. (2012). An overview of passive cooling techniques in buildings: Design concepts and architectural interventions. *Civil Engineering & Architecture, 55*, 84–97.

Kolokotroni, M., & Aronis, A. (1999). Cooling-energy reduction in air-conditioned offices by using night ventilation. *Applied Energy, 63*, 241–253.

Liang, X., & Huang, L. (2005). Optimum HVAC design suitable for the middle and small-scaled office buildings in South China with Sizzard climate. *Contamination Control & Air-Conditioning Technology, 4*, 61–63.

Liu, Y., Xing, Y., Zheng, W., & Li, N. (2015). Thermal performance optimizing combination of residential peripheral structure in moist heat zone. *Architecture Technology, 46*(5), 433–436.

Malone, A. (2010). *Best green project.* New York: McGraw Hill Construction.

Parker, D. S., Fairey, P. W., & McIlvaine, J. E. R. (1997). Energy efficient office building design for Florida's hot and humid climate. *ASHRAE Journal, 4*, 49–57.

Pérez-Lombard, L., Ortiz, J., & Pout, C. (2008). A review on buildings energy consumption information. *Energy and Buildings, 40*, 394–398.

Shui, B., & Li, J. (2012). (eds.). *Building EE policies in China, global building performance network*, Status Report. American Council for an Energy-Efficient Economy (ACEEE), Washington, D.C.

Sun, Y., Zhang, W., & Zhang, C., (2014). *Preliminary study on natural ventilation for hospital building in hot and humid regions.* 30th international plea conference, 16–18 December 2014, CEPT University, Ahmedabad.

USEIA. (2010). *International energy outlook 2010.* Washington, DC: U.S. Energy Information Administration, Office of Integrated Analysis and Forecasting.

Wang, S. K. (2000). *Handbook of air conditioning and refrigeration.* New York: McGraw-Hill Education.

Wang, J. (2011). Several problems in the research of green building design: Focused on hot and humid area. 绿色建筑设计研究中的几个问题——以湿热地区为例. *Huazhong Architecture,* 73–74. Available at: http://www.xueshu.com/hzjz/201104/12738453.html. Accessed on 21-Jan-2017.

WBCSD. (2007). *Energy efficiency in buildings, business realities and opportunities.* The world business council for sustainable development.

Xie, H. & Liu, X. (2006). Building thermal environment control of housing in tropical areas. 湿热地区的住宅建筑热环境控制. *Housing Science, 9*(10), 42–44.

Yang, L., & LI, Y. (2008). Cooling load reduction by using thermal mass and night ventilation. *Energy and Buildings, 40,* 2052–2058.

Yilmaz, Z. (2007). Evaluation of energy efficient design strategies for different climatic zones: Comparison of thermal performance of buildings in temperate-humid and hot-dry climate. *Energy and Buildings, 39,* 306–316.

Zain, Z. M., TAIB, M. N., & BAKI, S. M. S. (2007). Hot and humid climate: Prospect for thermal comfort in residential building. *Desalination, 209,* 261–268.

Zuo, J., & ZHAO, Z.-Y. (2014). Green building research – Current status and future agenda: A review. *Renewable and Sustainable Energy Reviews, 30,* 271–281.

CHAPTER 4

In Tune with Nature: A Low Energy Office Building in Tropical Colombo

Upendra Rajapaksha

Introduction

Nikini Automation is a multi-level medium size urban office building completed in 2008. In tune with nature, the office building harmonises its three-dimensional geometry with its microclimate as a solar defensive building form for heat gain control and thus energy conservation. Both its form and the placing on the site contribute to a dynamic feel. The open podium at the entrance subtly connects the building to the road creating a forecourt. The forecourt combined with the building's cubic form and sleek horizontal lines generate a lively atmosphere for its occupants.

The building is located in a tropical urban setting in Colombo City, Sri Lanka. Colombo is located at latitude 6.5 °N with an average temperature of 27 °C and daytime maxima reaching 33–34 °C. Nikini Automation sets an innovative design trend for environmentally sustainable architecture in Sri Lanka. The design illustrates climate control plus engineering systems for on-site energy and efficient integrated automation technology. The climate control interventions aim at reducing heat in and around the building and heat transfer from outside to inside. The architectural design theme centres around a daylight-sensitive layout but solar-defensive built

U. Rajapaksha (✉)
University of Moratuwa, Moratuwa, Sri Lanka

A. Cheshmehzangi, C. Butters (eds.), *Designing Cooler Cities*,
Palgrave Series in Asia and Pacific Studies,
https://doi.org/10.1007/978-981-10-6638-2_4

form, including a lightweight double skin to the east and west facades, reducing cooling loads and then meeting the balance of demand through building-integrated solar energy.

Challenges with the Climate in Colombo

Colombo is hot and humid all year round. The potential for interior overheating was considered as a main challenge at the initial design stage and possible interventions were considered. The site building is west-facing in a busy suburb, making the 'building–climate' interplay more challenging.

Hot-humid tropical climates are found in the region extending 15 ° north and south of the equator. Sri Lanka has this climate and little seasonal variation in temperature (Figs. 4.1 and 4.2).

Mean monthly temperatures range from 27 °C in November to 30 °C in April; relative humidity varies from 70% to 80% during a typical year. The daily temperature range in the dry season (September to November and March to May) is 7–8 °C.

The main characteristic of hot-humid climates, from the human comfort and building design viewpoint, is the combination of high temperature and high humidity. Givoni's psychrometric chart (Fig. 4.2) depicts indoor thermal conditions of buildings relative to outdoor conditions and highlights increased air movement to improve indoor thermal comfort. However, ventilation can create a heat gain in buildings in hot climates; and diffuse radiation is high in these climates too. This results in a conflict between attempts to control overheating and the role of natural ventilation. Thus, reducing the radiant heat in the incoming air is a high priority in naturally ventilated buildings. Controlling radiant heat at the microclimatic level, just outside a building, by good shading can reduce the heat load on the building.

The work of Humphreys (1978) and its continuation by Auliciems and Szokolay (1997) gives the optimum neutral temperature for human comfort as a function of monthly mean outdoor temperature. The neutral temperatures for Colombo are within the range 27.3 °C to 28.5 °C. A closer look at the ambient conditions during a typical year reveals that daytime temperatures remain above the extended upper limits of dry bulb temperature for most daytime hours (Fig. 4.1). Therefore, a strategy to reduce the daytime maximum indoor air temperature is the key in a climate responsive design approach.

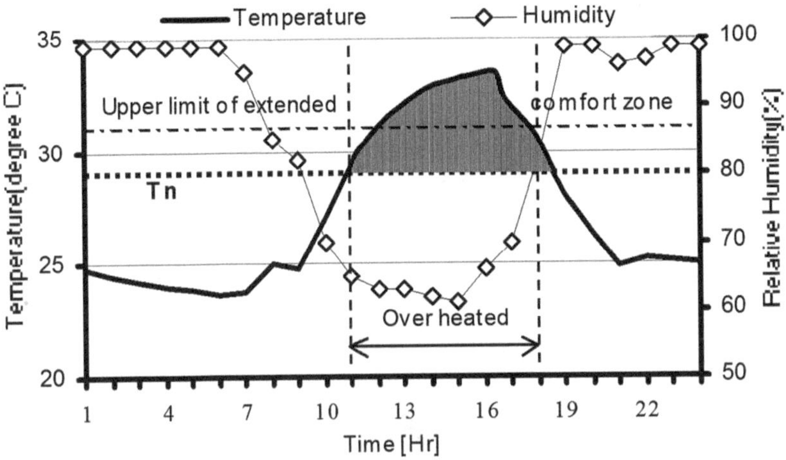

Fig. 4.1 Typical daily pattern of climate in respect to thermal comfort need in Colombo (Source: The Authors)

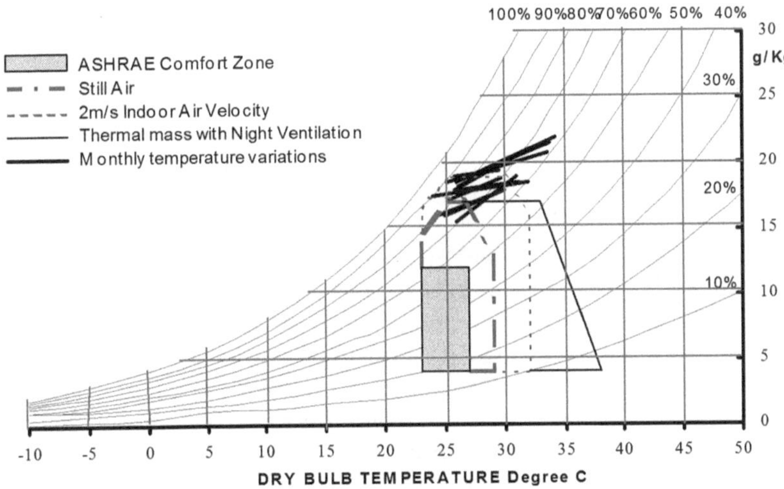

Fig. 4.2 The psychrometric chart applied to Colombo, Sri Lanka, 6.5 °N and 15 m above sea level (Source: Adapted from Givoni 1991)

A climate-responsive design approach in the tropics focuses on heat gain control and heat dissipation in order to improve the indoor thermal comfort, with low or nil energy consumption. The psychrometric chart (Fig. 4.2) indicates that cooler night ventilation can help to increase the heat sink effect of thermal mass and then contribute to reduce the elevation of indoor air temperature the following day, provided the building is closed to airflow from outside. However, the efficiency of this technique (thermal mass and night ventilation) is limited due to the small diurnal temperature range in tropical climates; whereas in moderate climates it is more effective due to the larger diurnal temperature differences (Rajapaksha 2003).

Limitations of Cooling Effect of Airflow

From the collective evidence of 11 different studies, Szokolay (1997, 2000) compared the cooling effect (extension of the comfort zone) of different air velocities and proposes a reasonable medium of cooling effect associated with air velocities up to 2 m/s. This work suggests that the cooling effect of air flow may diminish above this velocity level. Similarly, the work of Givoni (1991) suggests that, in the absence of solar or internal heat gains, the indoor air temperature closely follows the outdoor level with daytime ventilation. It also confirms that an air velocity of 1 m/s is capable of extending upper limits of acceptable dry bulb temperatures by 3.7 °C under warm humid conditions, in a situation where occupants wear light clothing and with low activity level. The evidence indicates a linear function of the cooling effect with air velocities up to 1 m/s but with diminishing effect above that.

Hence, indoor air movement of 1 m/s can extend the upper limit of the comfort zone in Colombo up to around 31.7 °C, without causing sensible thermal discomfort during summer. At the highest level of applicability, an air velocity of 1.5 m/s can extend the comfort zone by up to 5 °C, that is to 33 °C. A closer look at ambient climatic conditions for a typical summer day in this climate reveals that that daytime dry bulb temperature remains around 33–34 °C or higher. Given daytime humidity around 80% and slower air velocities, often no more than 1.0 m/s, there is a need for added indoor cooling. The Nikini building is designed for a conditioned indoor environment but is designed so that need for mechanical air conditioning is kept to a minimum.

A recent study in 2016 for Colombo (Jayathilake and Rajapaksha 2016) shows that nearly 90 % of current multi-level office buildings overheat during the daytime. Compact and core-dependent forms are designed to increase spatial and structural efficiency but are not designed with climate in mind. Although in the tropics the imperative is to prevent heat gain, and promote heat loss, office buildings in Colombo are very energy intensive due to poor climate responsiveness of the architectural design. This investigation revealed that 'low levels of building-climate interplay' result in susceptibility to overheated microclimates just outside the buildings and indoor overheating, from 3 °C to 4 °C above ambient levels, which resulted in a 30–50 % increase in the space cooling energy load.

Energy for building operation in Sri Lanka, in particular for offices, has been increasing for years; The Sri Lanka Sustainable Energy Authority (SLSEA) of the Ministry of Environment states that the average in typical offices in urban areas is as high as 260 kWh/m^2/year. Therefore, there is a high need for improving demand side energy efficiency in office buildings. Energy-efficient building design—the architecture—is considered as a first order priority, if the integration of second- and third-- order interventions such as PV panels is to be successful (Rajapaksha et al. 2015). The Nikini building highlights this approach to climate responsive design.

Scope and Design

In order to achieve a reduction in both energy use and its associated carbon emissions, an integrated 'design-plus-system' approach is required. This focuses on the various components of energy use in buildings. Interventions based on standards in the context of bioclimatic design have been considered as a priority for sustainable operation of the building. Moreover, the floor area of 948 m^2 comprises mainly typical commercial office functions, with a basement, ground floor and two upper floors. The top floor houses a residence and the covered rooftop is a solar deck for recreational activities. The west-facing site has a natural slope, which has been retained and used to create a sunken courtyard in front of the building.

The compact, envelope-dependent form of Nikini was designed to control environmental heat gain. For benchmarking thermal and energy performance of this office building, the following three priorities were defined: (a) Integration of architectural design interventions for climate response as the first order priority during the initial sketch design stage—this helps to increase the efficiency of the thermal performance of the building and reduce the demand for space cooling and artificial lighting; (b) Integration of high efficiency engineering systems as the second order

Fig. 4.3 Left: The aerial view of Nikini building with the roof PV panels and shading devices. Right: Front elevation of Nikini building (Source: The Authors)

priority; and (c) Integration of renewable energy generation with the building design as the third order decision (Fig. 4.3).

As a result of the above decisions the building maintains 2–3 °C lower indoor temperatures than the outside microclimate in non-air-conditioned mode. The main specific design components of the building are:

- Sunken courtyard in the section on the west, which creates a shaded microclimate in front of the building, reducing the development of radiation at ground level on the western side of the building facing a busy road with heavy traffic flow throughout the day – also mitigating anthropogenic heat from vehicles, another contributor to outdoor heat;
- Cantilevered floor plates on the building, shading the windows of the floor levels below;
- A double skin envelope with automated solar-sensitive lightweight horizontal louvres providing buffer zones for solar defence on east and west facades, in the morning and afternoon respectively;

- Daylight-sensitive building form with both plan and sectional geometry designed to optimise daylight without environmental heat gain;
- A solar roof as an 'umbrella' shading the upper floor and providing a power plant at the site;
- A minimalist approach to interiors, with daylight penetrating through the internal partitions and with healthy timber furniture.

The double skin façade of Nikini is a priority and an interpretation of the verandah concept in order to create shading around the building; particularly, on the façades facing direct sun, the east and west. In colder climates, the fairly recent introduction of systems with double façades serves several purposes, mainly related—but not only—to energy. A well-designed double façade in temperate climates serves to prevent wind chill, reduce heat losses and preheat incoming ventilation air, as well as providing significant noise reduction in urban contexts. In a hot climate, on the other hand, the principal function is to avoid unwanted heat gain by reducing solar incidence on the building facades. The function of reducing noise is important too. The louvre elements in the Nikini building also direct breezes on to the space within the double skin and help to avoid stagnation of hot air, thus, contributing to less heat transfer through windows. Furthermore, the double façade elements are of very light materials, and thus, add little to the carbon footprint of the building.

Air temperature readings taken in the sunken courtyard and buffer zones in the double skin envelope during hottest periods show the control of environmental loads by sectional and envelope characteristics (Fig. 4.4). Results from Hobos found that air temperatures inside the buffer zone created by the double skin just outside the glass windows are around 28–29 °C during peak daytime. This is 3–5 °C lower than the external microclimate. East façade readings indicate a reduction of 3 °C in the morning and 5 °C in the afternoon.

The façade louvres are adjustable according to solar angle. Whilst maximising solar protection, they simultaneously reflect dayight into the building, which is spread by reflection off the white ceilings, providing excellent indoor daylighting. Interior partitions are mainly of glass, thus again, maximising the spread of daylight inside. Here again, one building element is designed to fulfil several environmental functions. This integration of functions is a key factor in reducing building costs in environmental architecture. The best example of integration in green architecture today is probably that of solar photovoltaic roof tiles; as opposed to earlier solar

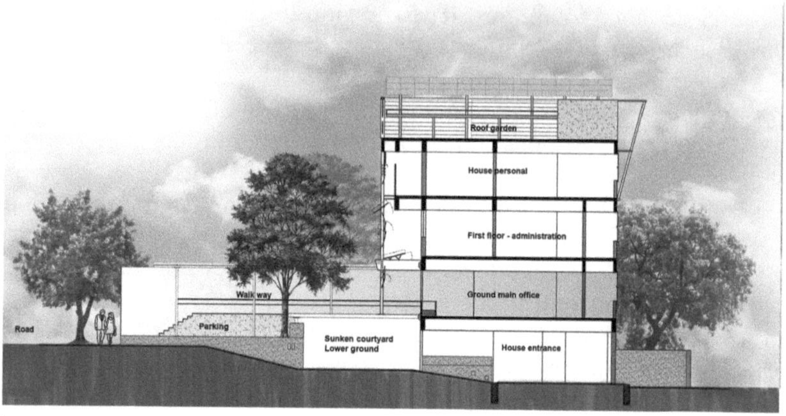

Fig. 4.4 The sectional drawing of Nikini building with spatial arrangement and landscaping features (Source: The Authors)

panels, which were additions on top of or beside buildings, these panels actually replace the roof tiles—the energy producing element becomes the roof itself.

In hot climates, flat roofs exposed to the sun are a major source of overheating of the top floor. The solution of a PV array on top of the Nikini building, thus, serves as double function of shading, for users of the roof terrace, and of converting the incident solar heat into useful electrical energy. A covered garden terrace on the roof also benefits from maximum local breezes and is a much-appreciated space for the users. In addition, the integration of the following efficient systems further reduces the environmental footprint:

- Motion sensitive active and task lighting systems;
- Variable Air Volume (VAV) air conditioning systems;
- Rainwater harvesting;
- Energy management system to monitor the performance of the building.

The third-order intervention of renewable energy generation (PV roof) reduces the final energy demand even further. This was a key objective, and it is now clear that of the total energy requirement of the building (about

60 kWh/m^2/year) nearly 50 kWh/m^2/year is covered by renewable energy generation on site with a grid-connected PV system. The roof has 180 photovoltaic panels which form a building-based power plant and produce 130 kWh per day on average. Since the energy demand is low, excess energy produced on sunny days is fed into the national grid. Total solar electricity generation since commissioning is nearly 300,000 kWh and total carbon saving is 174,304 kg (at 0.6 kg/kWh).

The Nikini building is the first of its kind in Sri Lanka. Operation with near-zero emission status started in 2008 with the approval of Ceylon Electricity Board (CEB). The CEB now promotes the net metering concept as a national target for Sri Lankan building owners. The SLSEA has recognised the project as the office building with the lowest energy footprint in an urban context. Nikini Automation highlights benchmarking standards for building design professionals for energy efficiency and integration of on-site renewable energy generation.

From Tradition to Sustainable Modernity

In addition to Szokolay and Givoni noted above, many authors, from Rudofsky (1970) to Hassan Fathy (1986) and Paul Oliver (1997), have pointed out and marvelled at the wisdom of traditional builders in adapting their shelter to local climate and resources. This chapter in fact recalls one of those authors, Ian McHarg (1995), whose classic book, *Design with Nature*, first appeared in 1969. The Nikini office building in Sri Lanka is an example of how modern, environmentally aware designers are reinvigorating evergreen principles—aided by modern scientific understanding, technology and design tools. One of the main goals is to create cool living and working environments in very demanding hot climates.

Whilst there are fundamental differences between hot-dry and hot-humid climates, the latter being the focus in this book, there are common principles. These can be found applied in different creative ways in hot climates all over the world. Over the past 20–30 years, hundreds of research projects and pilot buildings have investigated, analysed, improved and tested many of these techniques. But they are still only rarely applied. A few of these traditional or vernacular principles for 'staying cool' can be mentioned briefly, providing a contextual and cultural background to the design of this modern building in Sri Lanka.

The two first principles for providing comfortable temperatures in hot climate environments, are solar protection and air movement. The first can be seen in the many ways of shading buildings, with orientation that minimises solar gain, ample roof overhangs, vegetation and other means. This principle is fully applied in the Nikini building, as described, but including new technical advances such as the double facade approach. So is the second: maximising cross-ventilation and using permeable facades as well as interior partitions to allow air flow. Modern research, often aided by detailed local wind data and computer modelling now enables us to select the placement, shape and construction of overhangs, balconies, windows and other elements so as to maximise air flows into and through buildings. An air speed of 1 m/s produces the same perceived comfort effect as lowering the temperature by around 3 degrees. This is particularly important in humid tropical conditions.

A third general principle in the hot-humid climates is to use lightweight materials, which do not store heat or, equally important, humidity. This principle does not apply in the hot-dry climates—such as in North Africa—where heavy materials can exploit the cooler night temperatures and keep the building cool the next day. In modern cities this is not always applicable given the use of concrete and masonry. Modern research however shows that thermal insulation of heavy buildings is useful in these tropical contexts, if applied appropriately.

A fourth general feature of vernacular climatic buildings is that they were controlled by humans. Many of the most effective and cheapest solutions—opening and closing shutters for example—are simple manual operations, determined by occupants from one hour to the next according to their perceived needs. These operations have now largely been automated. Yet as we note elsewhere, control by the users of their indoor environment is a key quality and also leads to higher satisfaction and lower energy use. One should therefore approach 'smart technology' with a dose of scepticism, to achieve a sensible balance between the very efficient technical means now available, and human participation. For, ultimately, technology is only an aid, and there cannot be a 'sustainable' world without intelligent and resource-conscious people. Energy behaviour is not least cultural and cannot be ignored; the colonial culture of European-style office clothing is probably alone still responsible for more cooling than the combined energy production of several developing countries.

Creating shading outside buildings has been one of the most common approaches in vernacular building forms in Sri Lanka. This is visible in traditional courtyard buildings that are often also adapted for wind-induced cross-ventilation. The courtyard layout acts as an air funnel. To avoid heat penetrating to the interiors, perimeter openings are protected with heavy shading, in most cases using wide roof eaves and intermediate spaces such as verandas. Vernacular buildings with courtyards and other hot-climate features minimise direct solar radiation. Verandas, transitional spaces from outside to inside and to the courtyard, are sometimes partitioned with lattices or blinds, as multi-functional spaces used for circulation, eating, sleeping, relaxing, entertaining as well as sheltering and shading.

In addition to modern knowledge about global climate, energy and ecosystems, what we also have today is far more detailed knowledge about local climatology as well as about human physiology and comfort–which varies within cultures and climates. In addition, some modern tools such as life cycle analysis (LCA) and computational fluid dynamics (CFD), mentioned often in this book, which allow designers to optimise solutions in a way traditional builders could not.

Many other specific traditional cooling techniques can be found, often particular to a specific region; from evaporative cooling and solar chimneys to humidity buffering or wind cowls, these too have been extensively researched and have been given modern applications. It can be noted that these vernacular solutions apply both on the micro level for individual buildings but can be seen at the meso level in the design of village patterns, and indeed on the macro level too in the climate responsive location and layout of the small human settlements that are the forerunners of today's large cities.

Naturally, not all of these vernacular approaches can be relevant today. For example, in a modern dense urban context it is not possible to select climatically favourable locations to build, nor is it possible to maximise air flows as one might be able to in a country village. Polluted city air is not desirable for natural ventilation. Trees cannot shade tall multi-storey buildings. Use of water bodies as cooling elements is very limited on small city plots, although a small cooling pond is indeed integrated into the design of the IBR building (Chap. 3).

The common goal of all the passive design solutions is to make a building as comfortable as possible, even in the hottest seasons, by intelligent and climate-responsive design, without the use of added technology with its associated capital and energy costs—for installation, running and maintenance—and the possibility of technical breakdowns. In very hot and especially in hot-humid climates, passive design solutions alone may not be sufficient to ensure modern comfort standards all year round. There is, therefore, the need for added technology, which is selected so as to be as energy efficient as possible and if possible powered by renewable sources. This again is addressed in the Nikini building.

Near Zero Emission Building Categorisation

On sunny days the building's thermal performance meets the Net Zero Carbon Dioxide emission targets. This is an example of a net zero emission building (NZEB) since the accounting of renewable and fossil fuel energy takes place on site. The arrangement with the utility company is a gross tariff system where renewable energy is offset against non-renewable. More importantly, calculations for source, cost and emissions have been made and are documented for the inherent benefits of the renewable energy fraction. Major achievements of this project, include, demonstrating carbon saving in building practice, and creating a strong awareness amongst professionals and decision makers. The building has contributed to setting standards for low energy practice in Sri Lanka.

The embodied energy of materials used in construction was a concern during design of Nikini. The use of concrete for the structural frame and masonry brick for the outer envelope contributed to minimise transport energy as they are available locally. The lightweight panels of the windows and external frame of the double skin envelope are easily fixed on the main frame and can be recycled if necessary. Timber furniture throughout contributes to emission-free indoor air; and no synthetic polymers (plastics) have been used in the building's interior.

Policy Support for Net Zero Emission Buildings

The future is promising for NZEB buildings in Sri Lanka with the Government embracing sustainable development (based on reports of Sustainable Development Knowledge Platform by the United Nations, available at: h ttp://sustainabledevelopment.un.org/memberstates.html). The Energy Conservation Fund was established in the mid-1980s to coordinate activities in the electricity sector. Development of an energy-efficient building code primarily focusing on engineered systems was introduced in the early 1990s (Wijayatunga et al. 2003), focusing on commercial buildings, which consume nearly 21 % of total electricity production. Energy performance standards for buildings have been in place since 2006 for commercial and office buildings. SLSEA is the primary institution that delivers building-related standards. Promoting energy efficiency and conservation has become a priority.

The National Energy Policy and Strategies (2008) of the Ministry of Power and Energy envisages the gradual increase of non-conventional renewable energy to meet demand which is growing at around 8–9 % annually (Deheragoda 2009). Increasing renewable-energy-based electricity in the grid is a key strategy in a broad-based energy mix to provide clean and affordable energy to all sectors whilst minimising environmental impacts and greenhouse gas emissions. Given the decreasing percentage contribution from hydropower sources, biomass, solar and wind have been identified as sources in this regard (Fernando 2005).

Conclusions

Climate-responsive building form and integration of on-site renewable energy with PV panels are the main aspects of innovation to reduce energy demand and provide cool living and working environments. The sectional form with double skin envelopes and shallow depth plan form are major interventions to the architecture. The building form minimises environmental heat gain, optimises daylight, and thus, reduces cooling as well as lighting demand. Supply of renewable electricity from the building's roof with net metering to the national grid is the first of its kind in Sri Lanka. The building has already set a practice guide on how to implement multi-level urban office environments. These strategies can be applied to other and more complex situations if the basic thermal performance characteristics are known.

References

Auliciems, A., & Szokolay, S. V. (1997). *Thermal comfort, PLEA notes.* Brisbane: The University of Queensland.

Deheragoda, C. K. M. (2009). Renewable energy development in Sri Lanka, with special reference to small hydropower, TECH, MONITOR, Nov- Dec 2009. Available at: www.techmonitor.net/tmimages/e/ee/09. Accessed 28 Aug 2014.

Fathy, H. (1986). *Natural energy and vernacular architecture.* Chicago: University of Chicago Press.

Fernando, W. J. L. S. (2005). *Sri Lanka energy sector development.* In Proceedings of the 2005 workshop on Sri Lanka energy day, world energy council executive assembly, Colombo, Sri Lanka, 31–45.

Givoni, B. (1991). Performance and applicability of passive and low-energy cooling systems. *Energy and Buildings, 17.* Elsevier Sequoia, 177–199.

Humphreys, M. H. (1978). Outdoor temperatures and comfort indoors. *Building Research and Practice, 6*(2), 92–105.

Jayathilake, K., & Rajapaksha, U. (2016). *Interactive architecture and contextual adaptability: Issues of energy sustainability in contemporary tall office buildings in Colombo – Sri Lanka.* In Proceedings of the 9th FARU international research conference, University of Moratuwa, Colombo, 354–365.

McHarg, I. (1995). *Design with nature, 25th anniversary edition.* NY, USA: Wiley.

Ministry of Environment, Sri Lanka. (2008). *National Energy Policy,* Colombo.

Oliver, P. (1997). *Encyclopedia of vernacular architecture of the world.* Cambridge/New York: Cambridge University Press.

Rajapaksha, U. (2003). *An exploration of courtyards for passive climate modification in non-domestic buildings in moderate climates.* PhD Thesis, The University of Queensland, Australia.

Rajapaksha, U., Rupasinghe, H., & Rajapaksha, I. (2015). Resolved duality: External double skin envelopes for energy sustainability of office buildings in the tropics. Proceedings of 31st international conference of passive and low energy architecture conference, September 9–11, 2015, Bologna, Italy.

Rudofsky, B. (1970). *Architecture without architects.* New York: Doubleday.

Szokolay, S. V. (1997). *Thermal comfort in the warm-humid tropics.* In Iroceedings of the Architectural Science Association (ANZAScA) conference, 7–11.

Szokolay, S. V. (2000). *Dilemmas of warm-humid climate house design: Heavy vs. lightweight + cooling effect of air movement architecture, city, environment.* In Proceedings of PLEA conference, Cambridge, UK, 144–49.

United Nations Sustainable Development Knowledge Platform. *Report documents for Sri Lanka,* by United Nations Department of Economic and Social Affairs, Available at http://sustainabledevelopment.un.org/memberstates.html. Accessed 17 May 2017.

Wijayatunga, P. D. C., Fernando, W. J. L. S., & Shrestha, R. M. (2003). Greenhouse gas emission mitigation in the Sri Lanka, power sector supply side and demand side options. *Energy Conservation and Management, 44,* 3247–3265.

CHAPTER 5

Towards Cooler Buildings: The Case of Thailand

Chris Butters and Pattana Rakkwamsuk

Introduction

One of the most important tasks facing hot-climate cities is cooling. The objective of the ELITH research program (Energy and Low-Income Tropical Housing, ELITH webpage 2013–2016) in which the editors of this book took part, was to reduce energy use and climate emissions in the built environment. With a particular focus on low-income housing in hot-climate developing countries, this included issues of sustainable design and city planning as well as energy systems and policy. In the area of policy, a key action, in all countries, is to develop standards, codes and building regulations for energy efficiency (and/or carbon emissions; both approaches may be adopted). This was addressed for the case of Thailand where our ELITH partner was the Joint Graduate School of Energy and Environment (JGSEE) in Bangkok (JGSEE webpage 2016). This chapter describes processes, methods and challenges involved in introducing and implementing energy-efficiency policies in national building regulations and practice. After

C. Butters (✉)
University of Warwick, Coventry, UK

P. Rakkwamsuk
King Mongkut's University of Technology Thonburi (KMUTT), Bangkok, Thailand

A. Cheshmehzangi, C. Butters (eds.), *Designing Cooler Cities*, Palgrave Series in Asia and Pacific Studies,
https://doi.org/10.1007/978-981-10-6638-2_5

describing very briefly the context of Thailand, its buildings, energy use, trends, institutions and existing or recent initiatives in energy efficiency, we review the JGSEE research. The role of local organisations and institutions is highlighted. In the last section, we discuss broader issues and the relevance of these experiences for other countries.

This project addressed the micro level of individual buildings as defined in our introduction. As illustrated throughout this book however, creating cooler cities also demands measures at the meso level of the planning and design of cities, and at the macro level of their overall energy systems. In hot-climate developing country cities, the major energy-related challenge is that of reducing the needs for space cooling.

JGSEE developed benchmarks and tools for energy-efficient housing in Thailand after extensive background research and studies of existing housing, regional building traditions, recent initiatives and energy trends in buildings in both rural and urban Thailand. These are the necessary preliminary stages in such a process. Surveys were also conducted of energy behaviour, thermal comfort and perceptions of indoor environment. Current and alternative construction materials were studied in order to identify ways to achieve reductions in embodied energy. Case studies were collected from similar climates. The relevant agencies and authorities, in particular the public bodies responsible for housing and for energy in Thailand, were brought into the research process, these being the institutions responsible for implementation as well as subsequent development of such measures.

Energy Efficient Buildings

In temperate and cold climate nations, such as in Europe, building regulations have in recent years adopted increasingly stringent energy requirements. These have recently been taking major strides forward due to two main factors: on the one hand the focus on climate change, and at the same time the emergence of extremely low energy 'passive house' type solutions. Whilst initially aimed at reducing the need for space heating, using increased insulation and similar measures, solutions now focus as well on the embodied energy/carbon in construction, on electricity consumption, and on a holistic life-cycle-based view of buildings. The added focus on materials is not least because when operational energy for heating or cooling has been reduced to a minimum, the embodied energy to construct the buildings becomes very significant, up to or even more than half of the total lifetime impacts. An additional focus of recent years has been on increased integration of renewable energy supplies, solar in particular, in building design. This may be solar thermal to provide hot water, photovoltaics to provide electricity, or both. The universal trend is now towards buildings with near zero impacts, variously defined with terms such as zero energy, carbon neutral or even carbon positive buildings.

Objectives in hot climates are the same, but many developing countries have as yet not yet achieved major advances and indeed some still have no energy efficiency requirements for buildings at all. As discussed in our book, this is often because knowhow and financial resources are lacking, as well as institutions capable of planning, let alone enforcing, new practices.

In all climates, there is broad agreement that 'passive' measures—designing so that the buildings need as little energy as possible—are the priority. Added 'active' technology is the second order priority. This order of priorities is illustrated in our case of the Nikini building in Sri Lanka (Chap. 4). Passive measures include intelligent and 'climate responsive' design that makes maximum use of natural energy flows such as cooling breezes, water bodies, daylighting, solar heat, ground source cooling and similar. The technological focus, such as developing more efficient cooling systems, is naturally important too.

The Context

Thailand is a country with ample resources and knowhow. Available statistics and data are quite extensive, which is not the case in some countries. Most of Thailand has a tropical (hot-humid) climate, but with significant regional variations and local building traditions; background studies, data and surveys are available on the ELITH web site. As in many tropical climates, a general feature of traditional buildings is lightweight construction that does not accumulate heat, with good cross-ventilation and solar protection. It can be noted that such buildings have an inherent advantage compared to cold climate buildings with their many layers of insulation, double glazing and weatherproofing: being much lighter, they typically have less materials use and thus lower embodied energy. Modern buildings however, particularly in urban areas, are now mostly of heavy masonry materials such as concrete. In cities, good cross-ventilation is more difficult to achieve, as is solar shading especially in tall buildings. Hence the need for space cooling increases. Given rising affluence, air conditioning is experiencing explosive growth of over 10 % in some cases (Chirarattananon et al. 2015b). The prevalence of other heat-producing appliances—for lighting, cooking and other amenities—is also increasing rapidly. And the urban heat island effect is further aggravated by increasing traffic.

There exists a serious concern in Thailand about the increasing climate emissions due to rapid growth based largely on fossil fuels, serving to boost the industrial economy. The urban population is now over 40 %. Policies for low-carbon development are part of the official framework in Thailand, formulated in Five Year Plans by the National Economic and Social Development Board (NESDB) (World Bank and NESDB, 2011). Thailand has been pursuing energy and climate policies quite actively. For instance, a new 2009-version of building energy codes was enacted. It enforces energy efficiency when retrofitting commercial buildings having floor areas of 2000 m^2 and above to comply with a set of minimum performance standards for the building envelope, lighting and air-conditioning systems and promoting more use of renewable energy resources. Of particular interest is the fact that Thailand implemented a very large programme of low-income housing for a number of years, the *Baan Eau Athorn* programme, which produced several hundred thousand homes, discussed further later. The designs drew on local vernacular solutions and, whilst they cannot be described as advanced energy designs, they do incorporate several climate-responsive features, such as good shading and increased cross ventilation (Fig. 5.1).

Stages of the Process

In approaching the task of improving building practice at a national level, four key stages can be discerned.

Stage 1: Data and Background Studies

The Thai research team conducted extensive surveys on designs, materials and types of housing in the various regions of the country. Necessary climatic information includes temperature, humidity, rainfall and wind. The team gathered a large amount of data on housing energy use, needs and trends, plus surveys on indoor conditions and energy behaviour. Daylighting was also studied (ELITH webpage 2016, publications T01, T02, T06.). In the case of Thailand, much of the required data are already available in national and other statistics, as well as previous architectural and historical studies of regional building traditions. In other developing countries, such as our two ELITH partner countries Uganda and Tanzania,

Fig. 5.1 Traditional house, northern Thailand (Source: Chirarattananon et al. 2015a; also available on ELITH website)

such data may be scarce and, where available, also unreliable, necessitating much more groundwork. Alongside the above, information was gathered from other countries on relevant experience and best practice. Once assembled, this compilation of knowledge from present and past formed the basis for work on future solutions.

Stage 2: Benchmarks and Tools

Energy-specific benchmarks, methodologies and software are available now in many countries. There is much work in this field not least in Asia. The research team studied a number of systems such as BREEAM (UK), LEED (USA) and CASBEE (Japan), as well as various building regulations, in order to develop suitable solutions for the Thai context. Requirements for thermal comfort and indoor environment are to some extent culturally specific; it is well known for example that the US ASHRAE standards are not always applicable in tropical contexts, where perceived comfort may vary considerably (Kwong et al. 2014). Similarly, international databases for embodied energy and carbon, such as ICE (Hammond and Jones 2011) may provide useful and in some cases sufficient approximations but are often not applicable since industrial production processes for materials as well as primary energy mixes vary widely in different countries. For example, aluminium produced in Norway with renewable hydropower will have far lower carbon emissions than aluminium produced in China using coal. Similarly, the efficiency of cement production in different factories may vary from 800 to over 2000 kWh per ton of cement. Here again, fresh research is often needed in order to provide solid enough base data. The Thai team devoted particular attention to developing the methodologies for calculating the thermal performance of buildings; in most countries there are now one or several accepted methods of calculating energy flows and requirements in buildings, which designers must use to submit building proposals to comply with the requirements for energy performance.

Stage 3: Testing and Pilot Projects

Whilst the ELITH work focused mainly on the above two stages, the relevant authorities were involved, these being principally the National Housing Authority (NHA) (see www.nha.co.th) and the Electricity Generating Authority of Thailand (EGAT) (see www.egat.co.th/en). Discussions were held around testing the proposals plus the need for a programme of pilot buildings. This stage also requires consultation and feedback from the building industry, as well as preparing them for coming requirements. It is common experience that establishing dialogue and constructive engagement with the building industry may be difficult. This tends to be culturally specific; it has for example been easier to achieve in the Nordic countries where a large degree of consensus and cooperation exists. During this stage, dissemination and capacity building for planners and public officials as well as amongst design and build professionals is also important. Results on energy performance from pilot buildings may require several years: as will an assessment of calculation methodologies, costs, durability and user response.

Stage 4: Implementation

All such processes require quite a long period of development, testing and subsequent refinement. The move towards more energy-efficient building practice has almost everywhere started through individual initiatives and pioneering, followed by voluntary guidelines and recommendations (Loewenstein 2009). When these have proved their worth, they are incorporated into official, binding standards and regulations. Public awareness and support follows, if slowly. The objective in Thailand was the development of standards and tools for the energy performance of housing in particular; once achieved for housing, these advances can then be adapted for other types of buildings, for example schools, offices, hotels and so on. This has been the case with the BREEAM system in the UK. Over time these standards themselves are improved and made stricter—based on what is achievable in building practice in the specific country. Given the right incentives, examples and demands, building practice in Europe has been shown to be amenable to quite rapid change.

The Thailand Case

In Thailand there are around 21 million households and, as in most countries, the number of persons per household is decreasing and is now around 3.2. Some 77 % of households live in detached dwellings. With increasing urbanisation the number of apartment type buildings is however on the rise. (Bureau of Social Statistics 2014). More than half of all dwellings are now made largely of heavy materials such as brick and concrete. The advantages of lightweight buildings in a hot climate have been documented extensively (Gerilla et al. 2007). The embodied carbon of Thai residential buildings has been researched (Aneksaen 2011) and there are national guidelines for carbon intensity of buildings (Thailand Greenhouse Gas Management Organization 2011). In the urban areas, not least in the capital Bangkok, the urban heat island effect is compounded by rapidly increasing vehicle traffic and this is leading to as much concern about health as about rising energy use for cooling in particular.

Whilst traditional houses in Thailand, as in other hot countries, demonstrate intelligent adaptation to nature and climate, modern buildings are increasingly reliant on technology in order to stay cool. Hot conditions affect the poorest households in particular. Thailand has however devoted great attention to the low-income sector and there are few remaining 'slum' type areas (Usavagovitwong 2012). In 2002, the government mandated the National Housing Authority (NHA) to provide a very large programme of 'Baan Eau Athorn (BEA)' low-cost housing. Given the low-income focus, space efficiency was a key concern, and reducing space use is indeed the first and best way to reduce overall resource use and environmental impacts. Four housing types were developed (*Figure, one or several types in collage*): detached houses, twin houses, row houses and four-storey condominium apartments. The sites are in various locations but with over half in the Bangkok area. The government subsidised about a quarter of the cost of each unit. Eventually nearly 300,000 units were built (National Housing Authority of Thailand 2015). The floor sizes range from 31 m^2 to around 48 m^2. These units have slightly lower or similar energy use to other Thai houses, and the embodied carbon is also similar to other mainly heavy buildings in Thailand, between 100 and 170 $kgCO_2e/m^2$ of which, again, around 90 % is from the concrete and steel components (Ibn-Mohammed et al. 2013)—whereas traditional lightweight Thai houses have much lower embodied carbon. For comparison, the figures in typical highly insulated passive houses in Europe can be well over 200 $kgCO_2e/m^2$ (Fig. 5.2).

This programme carries useful lessons. On the one hand, developing functional housing designs with modest space needs and costs is a natural goal. On the other, these houses present a fairly similar picture to conventional building practice in Thailand as regards both operational and embodied energy or carbon. They were developed prior to today's focus on climate and energy. Whilst they do incorporate some features of passive cooling, these can be enhanced, and the embodied carbon in particular can be greatly reduced through the use of alternative materials, not least by incorporating low carbon concretes and more lightweight components

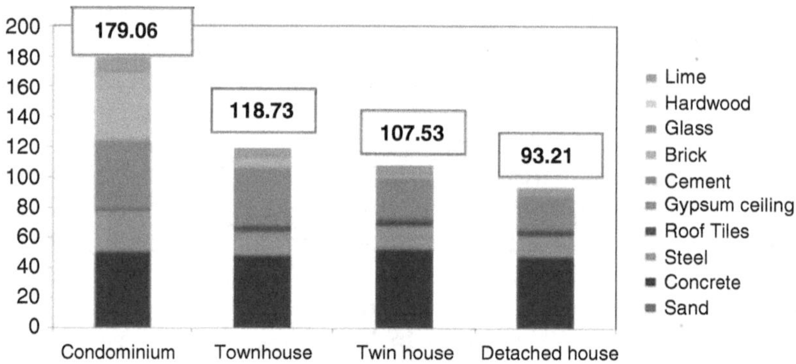

Fig. 5.2 Greenhouse gas emissions $kgCO_2e/m^2$ for four recent housing types in Thailand (Source: Loyprakhon et al. 2014; also available on ELITH website)

(Sirorangsee 2015). Since the programme included four different housing types it also offers an opportunity to develop benchmarks for cooling performance targets in typical housing types. This would be in line with the European case, where very low energy houses are expected to require for example a total 60 or 80 kWh/m^2.year, depending on country climate, of which only a small part (15 kWh/m^2.year in the 'passivhaus' case) is for space heating or cooling; and with similar benchmark figures having been developed and tested for other categories of buildings such as schools, offices, hospitals and so on. This then provides the basis for national energy requirements in the building codes.

The NHA was originally established in 1973 with the goal of alleviating the shortage of housing for low income groups, including management and financial assistance. Another key recent initiative of the NHA is the 'Floating Village' project (estudioOCA 2016), which has a strong focus on vegetation, native plants, bioengineering, community gardens and flooding, a major issue here as in other Asian cities, as well as a strong social and community focus. Such projects address our 'meso' level and are also very much part of a strategy for cooler cities.

The next step by the JGSEE team was to develop the calculation methodology for energy performance. Although lightweight construction is generally beneficial in hot climates, urban buildings are increasingly of heavy materials. Research including by ELITH team members has studied this more closely, and indicates that heavy constructions can in some cases be quite favourable. In particular it was found that 'thick walls may beneficially help delay heat gain for spaces that are used during daytime, the same effect increases cooling load for residential spaces that are used during night time' (Chirarattananon et al. 2012). This research is a significant pointer to recommendations for future building practice in order to reduce cooling energy needs. For the specific case of Thai housing, for example, where these are to be constructed in heavy materials it points towards the use of some insulation on the interior of external walls. Its interest reaches beyond that of housing too, since it illustrates the need to develop benchmarks that are specific for different functions of buildings—in this case showing how the case of daytime use as opposed to bedrooms differs, and that of offices from the residential case.

Similarly, the JGSEE team researched the energy consumption for cooling that results from excessive use of glass or unshaded windows. For example 'reduction in annual cooling energy consumption of 5 % to 15 % associated with reduction of solar heat input by 80 % to 90 % have been reported depending on the area and location of fenestration' (Chirarattananon and Hien 2011). This again compares with energy-efficiency guidelines in cold climates, which may recommend maximum desirable areas of glass in order to achieve good energy efficiency—although in the cold climate case this is to reduce heat loss as opposed to unwanted heat gain.

This example further serves to highlight an important point in building codes. Areas of glass, or insulation thicknesses, are not mandated in detail because different approaches can work. Given other design strategies even a glass building can, if with difficulty, be made energy efficient. European building regulations therefore now follow the general rule that they avoid specific requirements. One may use different solutions as long as the *overall* energy performance is achieved. This is extremely important in order to avoid unnecessarily detailed regulations—as well as to allow flexibility for new, creative solutions.

Other factors researched included heat gains through roofs and the influence of internal heat sources such as lighting, cooking and other appliances, all of which have an impact on cooling needs. A further, complex area of study is that of air-conditioned versus naturally ventilated buildings, on which there exists a large body of research. The Thai team thus investigated a broad spectrum of current knowledge, on building envelope performance in particular, as well as different tools and methodologies for calculating and simulating heat loss and energy flows (Fig. 5.3).

In their reports the Thai team recommended the rapid introduction of voluntary guidelines for housing energy performance standards. This too is in line with the European process, where a period of adaptation and experience is necessary before it is opportune to introduce obligatory, binding regulations. The authorities in any country will naturally be reluctant to introduce obligatory measures until a system has been proven to work and to not entail unreasonable extra costs. This again is a question of creating a process that is as rapid as possible yet pragmatic. Experience shows that the building industry is generally reluctant to make changes, and this opposition must be overcome by a combination of 'carrots and sticks'—incentives and restrictions. In Thailand there is already a statutory building energy code (BEC) for commercial buildings, with a calculation procedure for the exterior envelope of offices, department stores and hotels. Such methods for commercial buildings, based on the OTTV method—overall thermal transfer value—have been found to be suitable in some ASEAN countries. The resulting OTTV proposal which was developed by JGSEE for Thai

Fig. 5.3 Test house for energy, indoor comfort, ventilation and daylighting, JGSEE Thailand (Source: Chris Butters)

housing includes a range of calculations and modelling considerations (Chirarattananon et al. 2015b); these are not discussed in depth here since we are primarily addressing the processes towards energy efficiency in order to reduce the needs for cooling.

The proposed OTTV for residential housing addresses the thermal performance of an envelope enclosing air-conditioning areas. The OTTV is a representation of cooling loads, summing up those due to conduction heat gain and stored heat in opaque walls, heat stored in the interior room mass due to absorbed transmitted solar radiation, and conduction heat gain through glazing and resultant stored heat in the interior room mass. These, in fact, depend to a very large extent on the thermal properties of envelope construction materials, building orientation, window-to-wall ratio and time of use. The aim of JGSEE is integrating the OTTV methodology into a process of rating the energy efficiency of residential housing. It has also been reaching out to collaborate with authorities such as NHA and the Electricity Generation Authority of Thailand (EGAT) in order to implement the rating scheme.

Connecting the Meso and Macro Levels

Whilst measures for the energy efficiency of individual buildings are very important, many of the best and cheapest measures may be found at the meso and macro levels. At the meso scale this means designing neighbourhoods and urban layouts so as to create good microclimates and minimise cooling needs; at the macro level it involves the energy planning choices. It is important that these three levels are considered in conjunction. For example, in inner city contexts the options for climate responsive design are limited—natural ventilation may even be undesirable given polluted city air—and in such cases, ensuring macro level district cooling energy may be the best option. By contrast, where district cooling is less feasible more attention should be given to the meso and micro levels (Fig. 5.4).

At the micro level of individual buildings, reducing cooling needs by at least half is not an ambitious goal and does not require major changes in construction practices. At the meso level of urban design and layout, climate-responsive design can considerably improve the urban microclimate, hence enhancing comfort as well as reducing city cooling needs. At the macro scale, a district cooling system can further reduce the overall energy and climate impacts of hot cities, and in particular the heat island effect. Whilst all three levels can be addressed in new city developments, the case of existing cities and buildings is far more demanding. Energy efficiency

Fig. 5.4 Typical street townhouses and multi-level traffic, Sukhumvit, Bangkok, Thailand (Source: Photo taken by Ali Cheshmehzangi)

regulations are now also increasingly applied as requirements when buildings are to undergo major modification or refurbishment. At the meso level, it is very demanding to modify an overall city layout of cities but many cities are making large efforts to improve urban climate, increase vegetation and reduce pollution. At the macro level, similarly, there are major efforts to make energy supplies more renewable as well as to introduce district energy systems. District heating networks are increasingly common in temperate climate cities; but district cooling is as yet less known or applied.

Relevance and Conclusions

There are many features of the Thai experience that have immediate relevance for other countries with similar climates, albeit with necessary local adaptations; as well as for other regions outside Asia that have a hot-humid climate. Processes towards energy efficiency are much debated worldwide and present many local complexities, well-illustrated in the case of Thailand by a major Thailand/World Bank programme for Energy Service Companies (ESCOs), which whilst successful did not achieve widespread penetration in the market or the financial sector (Hansen 2009). There may be a need for several more programmes and system trials before the best options to achieve energy efficiency are identified. Nevertheless, specific standards as well as tools are broadly similar in similar climates. This process has seen considerable success in Europe, where the 'passivhaus' standards and specific building details, developed in Germany and Austria, were adopted as basis for developing national standards, codes and details in other European countries within a short period of time (Casals 2006). It may be added that European experience has shown that introducing very considerable energy efficiency improvements *does not* lead to large increases in building costs; of the order of just a few percent.

The challenge of promoting energy efficiency and disseminating it widely is first and foremost one of processes; the technical solutions exist. Thailand provides a useful example in this field. Cooling accounts for over 40 % of energy use in Thai households today; it is one of the major challenges that are receiving much attention and are being addressed in Thai policy (APEC 2010).The importance of cooling cannot be overstated when one considers that many other developing countries have as yet very little air-conditioning. Since this is likely to increase very rapidly, there is in many countries a very large opportunity for preventative action.

Cooling is primarily a human issue of health and comfort. One may add economic productivity, which, as is well documented, diminishes in hot environments, both factories and offices. The advantages of cooler cities are thus manifold. Almost all the most rapidly expanding cities are those of the hot-climate developing countries. And, whereas the rich in those countries can afford energy amenities such as air-conditioning, the growing heat island effect affects the poor most of all. Energy efficiency and cooling can be and needs to be addressed at all three levels discussed in this book; and it is imperative to see the three levels in conjunction in order to find optimal solutions.

References

Aneksaen, M. N. (2011). *Carbon intensity of residential buildings in Thailand.* Thonburi, Thailand: School of Energy, Environment and Materials, King Mongkut's University of Technology Thonburi.

Asia-Pacific Economic Cooperation (APEC). (2010, March). *Peer review of energy efficiency in Thailand.* Final report.

Bureau of Social Statistics. (2014). *Major findings on the 2013 household energy consumption.* Bangkok, Thailand: National Statistical Office, Ministry of Information and Communication Technology, Royal Thai Government.

Casals, X. G. (2006). Analysis of building energy regulation and certification in Europe: Their role, limitations and differences. *Energy and Buildings, 38*, 381–392.

Chirarattananon, S., & Hien, V. D. (2011). Thermal performance and cost effectiveness of massive walls under Thai climate. *Energy and Buildings, 43*(7), 1655–1662.

Chirarattananon, S., Hien, V. D., & Tummu, P. (2012). Thermal performance and cost effectiveness of wall insulation under Thai climate. *Energy and Buildings, 45*, 82–90.

Chirarattananon, S., Chiarakorn, S., Rakkwamsuk, P., et al. (2015a). *A survey of housing features and thermal comfort of medium and low income earners in Thailand.* JGSEE/KMUTT, Thonburi Thailand. Online publication T01. http://www2.warwick.ac.uk/fac/sci/eng/elith/publications/

Chirarattananon, S., Rakkwamsuk, P., et al. (2015b) *Development of energy performance indicators of residential building envelope.* KMUTT/ELITH 2015, Online publication T08. http://www2.warwick.ac.uk/fac/sci/eng/elith/publications/

Electricity Generation Authority of Thailand, EGAT. www.egat.co.th/en

ELITH webpage. (2013–2016). Energy and Low Income Tropical Housing, a 3-year research program funded by EPSRC (UK, lead partner Warwick University), 2013–16. http://www2.warwick.ac.uk/fac/sci/eng/elith/about/

ELITH webpage. (2016). Publications: T01, T02, T06. http://www2.warwick.ac.uk/fac/sci/eng/elith/about/

estudioOCA. (2016) NHA floating village, Bangkok. http://worldlandscapearchitect.com/nha-floating-village-bangkok-thailand-ssoca/#.WQxLe8YlFPY. Accessed 28 Apr 2017.

Gerilla, G. P., Teknomo, K., & Hokao, K. (2007). An environmental assessment of wood and steel reinforced concrete housing construction. *Building and Environment, 42*, 2778–2784.

Hammond, G., and Jones, C. (2011). *Inventory of carbon & energy (ICE) version 2.0*. Bath, UK: University of Bath.

Hansen, S. J. (2009). *ESCOs around the world: Lessons learnt in 49 countries*. USA: Fairmont Press.

Ibn-Mohammed, T., Greenough, R., Taylor, S., Ozawa-Meida, L., & Acquaye, A. (2013). Operational vs. embodied emissions in buildings – A review of current trends. *Energy and Buildings, 66*, 232–245.

JGSEE webpage. (2016). The Joint Graduate School of Energy and Environment, Thonburi: ELITH Team Surapong Chirarattananon (Principal Investigator of Thailand team), Pattana Rakkwamsuk, Siriluk Chiarakorn, Pipat Chaiwiwatworakul, Surawut Chuangchote, Preecha Tummu, Hathairat Loyprakhon, Bupphapan Puangbuppha. http://www.jgsee.kmutt.ac.th

Kwong, Q. J., Adam, N. M., & Sahari, B. B. (2014). Thermal comfort assessment and potential for energy efficiency enhancement in modern tropical buildings: A review. *Energy and Buildings, 68*, 547–557.

Loewenstein, O. (2009). *What Vorarlberg did next*. UK: Fourth Door review.

Loyprakhon, H., Chiarakorn, S., Rakkwamsuk, P. et al. (2014). *Greenhouse gas emissions from low income house construction: Case study in Baan Eua-Arthorn project in Thailand*. 5th international conference on sustainable energy and environment (SEE 2014), 19–21 November 2014, Bangkok, Thailand, online publication T04. http://www2.warwick.ac.uk/fac/sci/eng/elith/publications/

National Housing Authority of Thailand. (2015). Baan Eua Athorn. https://www.nha.co.th/view

Sirorangsee, S. (2015). *National housing authority of Thailand and urban nexus initiatives*. www.unescap.org/sites/.../Session%202-NHA%20GIZ%2017%20jUNE%202015.pdf. Downloaded 02 Sept 2016.

Thailand Greenhouse Gas Management Organization. (2011). *The national guideline carbon footprinting for local administration*. Thailand Greenhouse Gas Management Organization: Bangkok.

Usavagovitwong, N. (2012). *Successful approaches to national slum upgrading and prevention, Thailand*. Working Paper Series No. 06, Center for Integrated Socio-Spatial Research, Silpakorn University.

World Bank and NESDB. (2011). Thailand: Clean Energy for Green Low-Carbon Growth. www.documents.worldbank.org/curated/en/.../662200WP0p12440e0Clean0Energy0all07.p. Accessed 24 May 2017.

PART II

Case Study Chapters: Meso Level

In this section we address the meso scale of designing cooler layouts and environments for hot-climate cities such as residential and inner city blocks. The first is an example of research from Singapore into the important cooling effect of green infrastructures, in particular the role of trees. The second, a design case, describes a clustered housing development in India, which highlights the combination of building forms and landscaping to create a cool and sociable community environment. The third chapter in this section is a study by the editors for a large and partly high-rise urban development in China, which aims to illustrate an approach that integrates all three levels of micro, meso and macro considerations in a concept master plan solution.

CHAPTER 6

Cooling with Green Infrastructures: The Influence of Trees on Thermal Conditions in Tropical Urban Parks

Yun Hye Hwang, Qin Jie Geraldine Lum, and Li Xuan Cherlyn Lim

Tropical Urban Parks

Situated near the equator with average temperatures between 25 °C and 31 °C and relative humidity of 70–80 % (National Environment Agency 2015b), Singapore struggles with the issue of heat. This factor, in addition to the steadily increasing temperatures accompanying rapid urbanisation (Fong and Ng 2012), increases the urban heat island effect (Nieuwolt 1966; Roth and Chow 2012). Under the national initiative 'City in a Garden', vegetated settings are a common sight across Singapore, with a large number of public urban parks. There are large-scale plans to boost recreation and biodiversity in the parks and to increase the quantity of, and accessibility to, key outdoor gathering and activity nodes up to 900 ha by 2020 (Ministry of National Development 2013). While green policies may increase functional

Y.H. Hwang (✉)
Department of Architecture, School of Design and Environment, Singapore, Singapore

Q.J.G. Lum • L.X.C. Lim
National University of Singapore, Singapore, Singapore

A. Cheshmehzangi, C. Butters (eds.), *Designing Cooler Cities*, Palgrave Series in Asia and Pacific Studies,
https://doi.org/10.1007/978-981-10-6638-2_6

opportunities and indirectly benefit city-scale thermal conditions, it is believed that the park users' overall experience can be further enhanced with the concurrent consideration of thermal aspects. Several studies have stressed the significance of thermal comfort in urban park settings; parks are considered representative outdoor spaces where people are exposed to heat during recreation and leisure (Mahmoud 2011), and their thermal assessments (positive or negative) of those spaces could impact their patterns of behaviour (Thorsson et al. 2004; Yang et al. 2013).

Of all types of vegetation, trees are one of the most critical landscape elements in reducing temperature. This reduction normally occurs by the trees intercepting direct short-wave solar radiation, and filtering light and radiation through their shade-providing canopies (Erell et al. 2011; Kotzen 2003) before reaching ground level. Furthermore, trees that generate relatively large shade with extensive leaves and complex root systems effectively contribute in cooling down the temperature by absorbing ground heat through water evaporation process (Shashua-Bar and Hoffman 2000) and maintaining soil moisture (Monteith 1981), whereas lawns and/or shrubs may not do the same when there is a lack of shade. Some studies on the cooling effect of trees and tree canopies in urban parks highlight how employing tree clusters is more efficient than individual trees (Shashua-Bar et al. 2010), as the effect of shading may be integrated with other landscape elements such as water bodies (Müller et al. 2014). Many researchers have shown that trees are an effective means of minimising heat gain for humans, urban surfaces and buildings and have been studied in terms of building energy use and savings (Akbari 1992; Balogun et al. 2014; Shahidan et al. 2010). In addition, tree shade may affect meso scale cooling within/beyond urban parks (Bowler et al. 2010).

Given the thermal benefits of trees, it is encouraging to observe that Singapore has an overall vegetation cover of 56 % (Yee et al. 2011). Local studies have shown that these areas are likely to improve thermal conditions at the macro-scale of the city (Chen and Wong 2006; Wong and Jusuf 2010) and reduce the indoor temperatures of buildings (Wong et al. 2002). At the micro-level, field measurements from urban canyons and parks in Singapore demonstrate that the presence of tree shade reduces outdoor temperatures (Hwang and Tan 2013) and, consequently, improves thermal conditions (Hwang et al. 2015). Given these promising findings, the Urban Redevelopment Authority of Singapore has initiated thermal studies at the macro-level, hoping to moderate urban heat in the larger city-state (Urban Redevelopment Authority 2012).

While these studies have reinforced a general understanding that the presence of trees and shade can enhance outdoor thermal conditions, there is still a lack of consideration of optimum quantities and qualities of tree shade in the urban park settings or of optimal planting arrangements or types of tree needed to achieve the required shade for the optimum thermal comfort. This chapter discusses hypotheses that an investigation of the micro-scale thermal and shade performance of single trees would be able to determine the influence of shade and tree characteristics on heat in a more general way. One of the goals is to determine reasonable design and management strategies for the optimisation of micro-scale thermal conditions that can be implemented with as few calculations and measurements as possible. The specific objectives are:

1. To determine the thermal performance of a range of size of tree canopies and tree shade densities;
2. To identify characteristics and planting configurations of trees achieving optimal shade conditions.

The second part of the chapter describes methods of quantifying thermal conditions and shade characteristics. It also explains the tree specimen selection and measurement procedures involved in the study. The results section compares thermal conditions, shade and tree characteristics to determine their relationship, laying the foundation for a list of design considerations in the discussion section. Limitations and possible areas for further study are described in the conclusions.

Measuring Thermal Performance of Trees

Measurement Variables

Trees are known to reduce radiation; however, unlike similar thermal studies of trees (Kotzen 2003; Shahidan et al. 2010), this study does not consider the radiation flux. Rather, the given focus is on thermal comfort. To understand and quantify how a tree and its shade can alter the thermal experience of a person standing or sitting under it, the study uses the Predicted Mean Vote (PMV) model of thermal comfort as a suitable and recognised index of human acceptance of thermal conditions. Initially developed by Fanger (1972) for indoor environments, the PMV model is referenced in thermal comfort standards (International Organization for

Standardization 1994) and has been applied in outdoor studies (Jendritzky and Nübler 1981; Yang et al. 2013). In one example, a survey of Thermal Sensation Vote (TSV) in outdoor urban spaces in Singapore concluded that about 80 % of people would be satisfied with a range of outdoor operative temperatures from 26.3 °C to 31.7 °C (Yang et al. 2013).

As a result, the study uses operative temperature to quantify thermal conditions calculated using air temperature, mean radiant temperature, and wind velocity. Mean radiant temperature is calculated based on measurements of globe temperature, an equation derived from ISO standards (International Organization for Standardization 1998). This process is conducted to account for the convective conditions in Singapore's outdoor tropical climate (Tan et al. 2013). In short, thermal conditions are quantified by measuring and calculating air temperature, globe temperature, and wind velocity. See Table 6.1.

The characteristics of trees were measured and recorded to relate each to the corresponding shade conditions produced. The size and height of the tree canopies were expected to have a direct relationship with the area of shade produced, while the canopy density, comprising branch density and leaf cover was expected to affect the amount of light filtered under the canopy (Brown and Gillespie 1995).

To ensure similar environmental conditions for the tree specimens, the study looked at urban parks in Singapore with relatively open spatial configurations and large flat lawn areas. It singled out solitary trees within open grass lawns, with each tree some distance from the other and few or no shrubs present to reduce variations in the trees' environment (Fig. 6.1). Ultimately, the survey selects six tree types representing the typical specimens in an urban park, Clementi Town Park and Jurong Central Park, basing selection on their size (small, medium or large) and canopy density (sparse or dense).

Furthermore, the tree specimens were selected to have similar canopy shapes to ensure a good basis for comparison. As shade quantity is inevitably affected by the size of a tree, the study selects a range of tree sizes for assessment: (a) small trees defined as 2–3 m tall, with a canopy width of about 2 m; (b) medium trees, 5–6 m tall, and with a canopy width of 4–6 m; (c) large trees, 10–11 m tall and with a canopy width above 10 m. The study also considers two types of canopy densities, namely 'sparse' and 'dense', to investigate the influence of shade quality on thermal conditions. Figure 6.2 demonstrates the selected six trees and their immediate surroundings.

Table 6.1 Thermal variables recorded and equipment specifications

Variables	*Equipment*	*Measurement range*	*Accuracy*
Air temperature	Testo 445, with 3-function probe (shielded air temperature probe)	−20–70 °C	±0.4 °C from 0 °C to 50 °C
Wind velocity	Testo 445, with 3-function probe (wind velocity probe)	0.0–10 m/s	±5% or 0.03 m/s whichever is greater
Globe temperature	Extech HT30 globe thermometer[a] (Ø = 40 mm)	0–80 °C	±2 °C
Lux	TES1335 digital light meter	0–400,000 lux	±3 % of reading ±0.5 digits

[a]Extech HT30 globe thermometer has been compared to HOBO Thermocouple Data Logger, UX100-014 M with Type-T Copper-Constantan thermocouple sensors and 40 mm diameter grey ping pong ball. Note that the same instruments eq. (2) was calibrated for Tan et al. (2013). The amount of variation of t_{mrt} value was within a ±2 °C difference under park conditions from 12 noon to 2 p.m

Fig. 6.1 View of a typical urban park in Singapore

Fig. 6.2 Views of selected tree specimens and their immediate surroundings. Scales vary (*SS pongamia pinnata* (W1.7 m H2.7 m), *SD gardinia carinata* (W1.8 m H2.2 m), *MS cratoxylum cochinchinense* (W4.6 m H5.4 m), *MD alstonia angustiloba* (W5.1 m H5.5 m), *LS pongamia pinnata* (W11.5 m H10.8 m), *LD samanea saman* (W12.1 m H10.9 m))

Field Observation Period, Procedure and Equipment

Measurements were carried out between 12 noon and 2 p.m. in the months of April to May in 2015, the warmest time of the year with one of the highest levels of solar irradiance (National Environment Agency 2015a). To account for day-to-day variations in weather, each tree was measured on three different days to obtain average readings for each point. Measurements were made only when shade was visible; that is on sunny days with no obstruction from rain or clouds.

The study mapped the shade cast by each tree specimen from 12 noon to 2 p.m. by demarcating points 1 m apart (0.5 m for small trees) north, south, east and west of each tree, starting at a distance of 1 m (0.5 m for small trees) from the trunk and stopping before the point at which no shade was cast

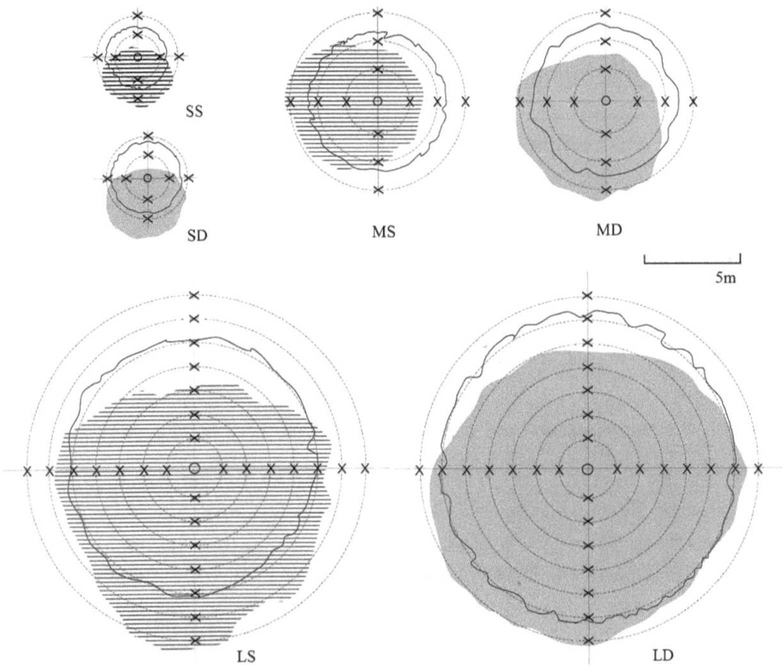

Fig. 6.3 Measurement points for trees, for shade cast from 12 to 2 p.m.; '*x*' represents a point where a temperature measurement was recorded

(Fig. 6.3). Once the measurement points were determined, the following measurements were made:

1. Thermal variables (Ta, Tg and Vel) and Lux were measured at the same time for each point, at the heights of 1.1 m and 1.5 m respectively (Mayer and Hoppe 1987).
2. Measurements of Ta and Tg were logged only when shade was distinct and after numbers on the equipment had stabilised before moving to the next point.
3. The process was executed over 3 days for each tree during the 2-hour period; the measurement time was consistent on all 3 days to avoid changes in the position of the shade in relation to the tree.

The size of the trees affected the number of required measurement points, with more points for the larger trees, so opting for a two-hour period limited the data collection to one large or medium tree or two small trees at a time. Thus, the entire observation period took a total of about 14 days to complete.

As four variables had to be recorded at the same time for up to 96 measurement points (8–28 points for each tree) within an hour, it was imperative to select portable industrial-standard equipment, The types of equipment and their function are listed in Table 6.1.

Factors That Affect Thermal Performance of Trees

Overall Thermal Performance: Size of Tree Canopy

Based on the findings, the average temperature T_o ranges widely in the shaded areas, from 37.7 °C to 44.3 °C, a difference of 6.6 °C. The overall T_o ranges from 39.8 °C to 47.6 °C; the former figure corresponds to the large dense tree and the latter to the small sparse tree, showing an increase of 7.8 °C. The large and medium trees performed better than the small ones. The differences between the thermal performances of the large and medium size canopy trees with similar densities were less significant, but the larger ones had a better thermal performance.

To verify the thermal performance, spot measurements were conducted for morning and evening periods. Morning and evening temperature records generally mirrored the results recorded in the afternoon (i.e., 12–2 p.m.). Table 6.2. shows the overall T_o performance of the selected six trees and shaded areas.

Light Filtration: Density of Tree Shade

The average T_o performance has noticeable differences that could be linked to the shade intensity. The average T_o of the large dense tree is 4.7 °C lower than the large sparse tree from noon to 2 p.m., 4.1 °C lower for the medium trees, and 2.0 °C lower for the small trees. The average T_o recorded in sparse shaded areas ranges from 42.1 °C to 44.3 °C, and the average T_o in the area of dense shade ranges from 37.7 °C to 40.7 °C. Thus, a higher shade intensity with lower light filtration achieves up to 6.6 °C reduction in T_o during the hottest period.

Table 6.2 Average T_O performance of the six trees and size of the shaded area

	T_o *(overall average)*	*Ave.* T_o *(shaded area) (°C)*	*Ave.* T_o *(unshaded area) (°C)*	*Shaded area* (m^2)
SS	47.6 (42.1[a])(43.3[b])(50.8[c])	44.3	51.8	3.09
SD	45.6 (40.3[a])(40.4[b])(50.6[c])	40.7	48.7	3.78
MS	45.4 (38.5[a])(40.1[b])(49.0[c])	43.5	48.1	20.75
LS	44.5 (37.8[a])(40.1[b])(49.2[c])	42.1	48.8	114.75
MD	41.3 (37.4[a])(38.0[b])(49.1[c])	39.4	46.8	25.22
LD	39.8 (36.8[a])(37.5[b])(50.4[c])	37.7	49.2	158.27

[a]Morning (8–10 am)
[b]Evening (4–6 pm)
[c]Ave. T_O of the control point ranged 49.0–50.8 °C in the afternoon (12–2 pm)

Next, shades of different intensities and corresponding thermal performances were compared for the entire day. Graph 6.1 indicates a positive linear relationship between the two variables; as the graph shows, increasing light filtration met with increasing operative temperatures in all cases. The lower spectrum of the graph contains six points; all six are medium or large trees measured in the morning or evening; and four have dense canopies. As the graph suggests, shade intensity values or the level of light filtration can be recognised as a gauge of thermal performance, based on the correlation of Lux levels and temperatures recorded. More specifically, the two large trees had average light infiltration values of 34.0 Klux for the sparser tree and 9.7 Klux for the tree with a denser canopy. Overall, for all six trees, a higher shade intensity (lower light filtration) of 50 % achieved average T_Os of 36.5–41.3 °C (see Graph 6.1).

Cool and Hot Points: Edge Effect and Sun Orientation

Cool points were identified from the thermal values recorded during the hottest time of the day. The coolest 11 points constituting the lowest 10 % of temperatures recorded ranges from 37.1 °C to 38.0 °C. All these values were identified for the large dense tree, whereas values for the other five trees exceeded 38 °C. Many of the coolest points were located in the middle of the shade, and cool points were generally located in the northwest direction in the afternoon, which corresponds to the orientation of the sun. The seven hottest points ranged from 51.0 °C to 51.9 °C. Most of these values were observed for the small trees regardless of their shade

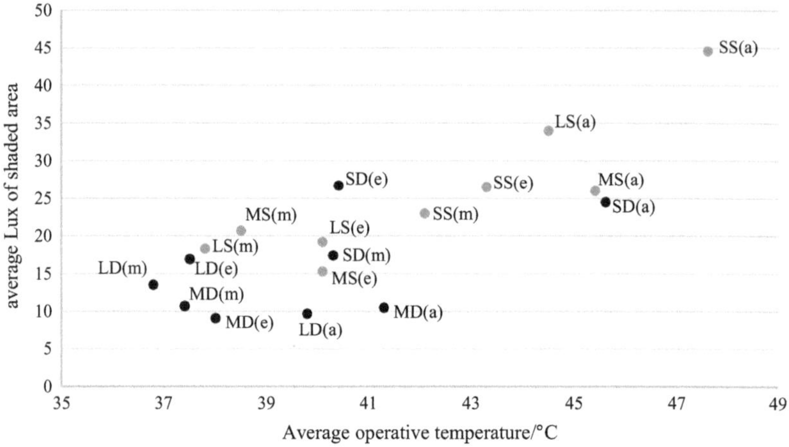

Graph 6.1 Operative temperature and light intensity in shade under canopy (Note: Darker points (•) are coordinates of a tree with a dense canopy, and lighter points (•) are coordinates of a tree with a sparse canopy. Points are labelled according to their name and the time-period when measurements were recorded, where *m* morning, *a* afternoon, *e* evening)

density, and all points were in the edge areas away from the shaded area offered by the large dense tree (Fig. 6.4). Thus, larger areas of shade are more effective at providing thermal comfort during the hottest time of the day than are smaller areas of shade.

Suggestions for Designing Cooler Parks

For hot environments such as tropical climates, the mitigation of solar and terrestrial radiation in landscape design is much more important than other microclimatic factors (Brown and Gillespie 1995; Emmanuel 2005). This study examined the thermal performance of trees in a tropical urban park, looking specifically at aspects of the shade characteristics of trees, including size and density. It found differences in the level of T_o performance amongst six trees of varying size and canopy density. The following findings can be applied to the planting and landscape design of tropical urban parks:

– An area of shade as small as 3.09 m^2 can reduce T_o by 3.2 °C; an increase of 25.22 m^2 may reduce T_o by another 4.6 °C and an increase

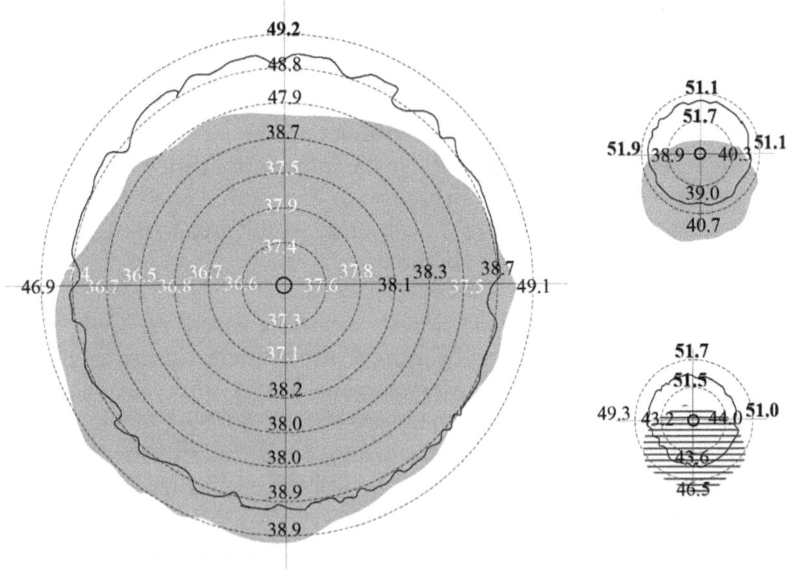

Fig. 6.4 Cool and hot points in surveyed trees on 12–2 p.m. The text in white highlights the position and operative temperatures of the 11 cool points (lowest 10% of temperatures recorded). The text in bold highlights the position of the 7 hottest points in small dense trees (top right image) and small sparse trees (bottom right image). Scales vary

of 158.26 m^2 means a reduction of up to 10.6 °C. Although the presence of low-intensity shade is better than the total absence of shade, higher shade intensities are recommended as they can reduce T_o by up to 4.7 °C. The findings can be applied to the selection of tree shade characteristics.

- Depending on context variables, the shape and growth of solitary trees can be altered by maintenance. Results suggest shade quality can be measured by lux. This finding may be useful in tree management, for example, to determine crown pruning and monitoring activities in parks.
- Cool points found by this study tended to be located within 5 m west of the trunk of the large dense tree, making this the ideal location of intended functional areas. The finding suggests large trees should be

preferred over small trees, and saplings should be planted a maximum of 5 m apart to achieve a preferred level of thermal comfort.
- Ideal locations to plant trees to create function areas may be determined by the orientation. The western orientation is preferable for functional areas at noon, but considering the movement of the sunlight throughout the day, it can be inferred that there will be more cool points in the west orientation in the morning and the southeast orientation in the evening. The finding may be used to design outdoor spaces for different usage by time.

As the thermal influence of solitary trees was assessed, the study cannot quantify the combined effects of groups or continuous stretches of trees; and these may lead to even lower temperatures. A survey of micro-scale thermal conditions in tropical urban parks in Singapore found temperatures were lower in the middle of relatively larger quantities of shade (Hwang et al. 2015). While quantification of the influence of larger areas of shade would be beneficial, the outdoor spaces surveyed by this study did not have a comparable range of varying shade quantities; thus, the effect should be quantified by further simulation studies.

Based on the results of the study, larger trees are recommended over smaller trees to optimise the outdoor thermal comfort. However, trees take time to mature, and it is not always possible to transplant mature trees because of difficulties in transport and establishment of the plant. Thus, the time taken for the trees to reach maturity and develop the desired characteristics should be factored into the design process. The planting distance from paths caused by the root growth of the trees should also be considered. Finally, it would be useful to explore the thermal perception of being in a group of trees, as well as bodily adjustment to 'flow'; for example, the experience of walking through continuous shade as compared to dashes of shade.

The study's findings should trigger further investigation of a variety of other tree forms so that a more comprehensive range of planting strategies may be recommended.

References

Akbari, H. (1992). *Cooling our communities: A guidebook on tree planting and light-colored surfacing*. Berkeley: Lawrence Berkeley Laboratory.

Balogun, A. A., Morakinyo, T. E., & Adegun, O. B. (2014). Effect of tree-shading on energy demand of two similar buildings. *Energy and Buildings, 81*, 305–315. https://doi.org/10.1016/j.enbuild.2014.05.046

Bowler, D. E., Buyung-Ali, L., Knight, T. M., & Pullin, A. S. (2010). Urban greening to cool towns and cities: A systematic review of the empirical evidence. *Landscape and Urban Planning, 97*(3), 147–155. https://doi.org/10.1016/j.landurbplan.2010.05.006

Brown, R. D., & Gillespie, T. J. (1995). *Microclimatic landscape design: Creating thermal comfort and energy efficiency.* New York: Wiley.

Chen, Y., & Wong, N. H. (2006). Thermal benefits of city parks. *Energy and Buildings, 38*(2), 105–120. https://doi.org/10.1016/j.enbuild.2005.04.003

Emmanuel, R. (2005). *An urban approach to climate-sensitive design: Strategies for the tropics.* New York: Spon Press.

Erell, E., Pearlmutter, D., & Williamson, T. (2011). *Urban microclimate: Designing the spaces between buildings.* London: Earthscan.

Fanger, P. O. (1972). *Thermal Comfort.* Düisseldorf/New York/London: McGraw Hill Book Company.

Fong, M., & Ng, L. K. (2012). *The weather and climate of Singapore* (2nd ed.). Singapore: Meteorological Service Singapore.

Hwang, Y. H., & Tan, Y. C. (2013). A study on greenery factors for cooler park connectors in Singapore. *The International Journal of the Constructed Environment, 3*(1), 1–13.

Hwang, Y. H., Lum, Q. J. G., & Chan, Y. K. D. (2015). Micro-scale thermal performance of tropical urban parks in Singapore. *Building and Environment, 94*, 467–476. https://doi.org/10.1016/j.buildenv.2015.10.003

International Organization for Standardization. (1994). *ISO 7730:1994 Moderate thermal environments: Determination of the PMV and PPD indices and specification of the conditions for thermal comfort.* Geneva: ISO.

ISO, I. (1998). *7726, Ergonomics of the thermal environment, instruments for measuring physical quantities.* Geneva: International Standard Organization.

Jendritzky, G., & Nübler, W. (1981). A model analysing the urban thermal environment in physiologically significant terms. *Meteorology and Atmospheric Physics, 29*, 313–326.

Kotzen, B. (2003). An investigation of shade under six different tree species of the Negev desert towards their potential use for enhancing micro-climatic conditions in landscape architectural development. *Journal of Arid Environments, 55*(2), 231–274. https://doi.org/10.1016/s0140-1963(03)00030-2

Mahmoud, A. H. A. (2011). Analysis of the microclimatic and human comfort conditions in an urban park in hot and arid regions. *Building and Environment, 46*(12), 2641–2656. https://doi.org/10.1016/j.buildenv.2011.06.025

Mayer, H., & Hoppe, P. (1987). Thermal comfort of man in different thermal environments. *Theoretical and Applied Climatology, 38*, 43–49.

Ministry of National Development. (2013). *Population white paper: A sustainable population for a dynamic Singapore.* Retrieved from Singapore http://www.ura.gov.sg/uol/publications/research-resources/plans-reports/Concept%20Plan%202011/land_use_plan_2013.aspx

Monteith, J. L. (1981). Evaporation and surface temperature. *Quarterly Journal of the Royal Meteorological Society, 107*(451), 1–27.

Müller, N., Kuttler, W., & Barlag, A.-B. (2014). Counteracting urban climate change: Adaptation measures and their effect on thermal comfort. *Theoretical and Applied Climatology, 115*(1), 243–257. https://doi.org/10.1007/s00704-013-0890-4

National Environment Agency. (2015a). *Daily values of meteorological elements in 2014.* Singapore: National Environmental Agency. Retrieved from http://www.nea.gov.sg/weather-climate/climate-information/weather-statistics

National Environment Agency. (2015b). *Local climatology.* Retrieved June 10, 2014, from http://www.nea.gov.sg/weather-climate/climate-information

Nieuwolt, S. (1966). The urban microclimate of Singapore. *The Journal of Tropical Geography, 22,* 30–37.

Roth, M., & Chow, W. T. L. (2012). A historical review and assessment of urban heat island research in Singapore. *Singapore Journal of Tropical Geography, 33*(3), 381–397. https://doi.org/10.1111/Sjtg.12003

Shahidan, M. F., Shariff, M. K. M., Jones, P., Salleh, E., & Abdullah, A. M. (2010). A comparison of Mesua Ferrea L. and Hura Crepitans L. for shade creation and radiation modification in improving thermal comfort. *Landscape and Urban Planning, 97*(3), 168–181. https://doi.org/10.1016/j.landurbplan.2010.05.008

Shashua-Bar, L., & Hoffman, M. E. (2000). Vegetation as a climatic component in the design of an urban street: An empirical model for predicting the cooling effect of urban green areas with trees. *Energy and Buildings, 31*(3), 221–235.

Shashua-Bar, L., Tsiros, I. X., & Hoffman, M. E. (2010). A modeling study for evaluating passive cooling scenarios in urban streets with trees. Case study: Athens, Greece. *Building and Environment, 45*(12), 2798–2807. https://doi.org/10.1016/j.buildenv.2010.06.008

Tan, C. L., Wong, N. H., & Jusuf, S. K. (2013). Outdoor mean radiant temperature estimation in the tropical urban environment. *Building and Environment, 64,* 118–129. https://doi.org/10.1016/j.buildenv.2013.03.012

Thorsson, S., Lindqvist, M., & Lindqvist, S. (2004). Thermal bioclimatic conditions and patterns of behaviour in an urban park in Goteborg, Sweden. *International Journal of Biometeorology, 48*(3), 149.

Urban Redevelopment Authority. (2012). Designing our city: Planning for a sustainable Singapore. In *skyline [Supplement].* Singapore: Urban Redevelopment Authority.

Wong, N. H., & Jusuf, S. K. (2010). Study on the microclimate condition along a green pedestrian canyon in Singapore. *Architectural Science Review, 53*(2), 196–212. https://doi.org/10.3763/asre.2009.0029

Wong, N. H., Chen, Y., Ong, C. L., & Sia, A. (2002). Investigation of thermal benefits of rooftop garden in the tropical environment. *Building and Environment, 38*, 261–270.

Yang, W., Wong, N. H., & Jusuf, S. K. (2013). Thermal comfort in outdoor urban spaces in Singapore. *Building and Environment, 59*, 426–435.

Yee, A. T. K., Corlett, R. T., Liew, S. C., & Tan, H. T. W. (2011). The vegetation of Singapore – An updated map. *Gardens' Bulletin Singapore, 63*(1&2), 205–212.

CHAPTER 7

Designing Residential Microclimates: Malhar Eco-Village in Bangalore, India

Ali Cheshmehzangi, Ayotunde Dawodu, and Amir Cheshmehzangi

Introduction

Around half of Mumbai's total population of 12 million today live in crowded and unhealthy informal settlements (National Sample Survey Organisation 2015). In India, as in many other developing countries, thermal conditions are often extremely poor, both in the inner city and especially in lower-income districts and slum-type informal settlements. This applies to indoor conditions as well as outdoors where people also live and work. Nevertheless, even amongst poorer groups, energy amenities including air conditioning (AC) are spreading very fast. Wang et al. (2013) note that the energy used for cooling globally could increase by 350 % by 2100. Studies in the United Arab Emirates suggest that cooling energy there would increase by 23.5 % in the coming decade. AC in India is the largest component of overall residential power consumption, and also correlates to the highest peaks in electricity demand; AC sales are predicted to

A. Cheshmehzangi (✉) • A. Dawodu
The University of Nottingham Ningbo China, Ningbo, China

A. Cheshmehzangi
AVA Consultants International Ltd, Derby, UK

A. Cheshmehzangi, C. Butters (eds.), *Designing Cooler Cities*,
Palgrave Series in Asia and Pacific Studies,
https://doi.org/10.1007/978-981-10-6638-2_7

increase by 15 % each year. The estimated 4 million new AC units installed in India by 2030 will lead to a requirement for 80–145 Gigawatt of additional energy (Phadke et al. 2014). Creating cooler cities, with limited resources, is very much a question of urban form and design. Compactness is a key issue. This chapter describes a residential neighbourhood in India that is designed to provide a comfortable microclimate with a minimum of energy or technical means.

Buildings are the interface between the indoor and outdoor environment. In hot climate cities, where cooling is a main need for human comfort, temperatures are rising, in addition to more frequent extreme heat events (Fell 2012). Population increase, the spread of energy amenities and vehicles, as well as global warming, all contribute to this. In addition, the urban population spends an estimated average of 85–90 % of time indoors (Tejero-González et al. 2016). This has especially severe ramification in developing regions such as in Africa and Asia where there are many countries with extremely hot climates; and even where people have sufficient resources to pay for energy, there is often also an energy gap where population growth exceeds the spread of electrification (Global Network of Energy for Sustainable Development 2007; Barnes and Foley 2004). The urban heat island (UHI) effect, noted throughout this book, is linked to high stress levels, morbidity and increased mortality; for instance, Paris, Melbourne and Moscow experienced increased mortality rates of 130 %, 62 %, and 60 % directly linked to extreme heat events (Norton et al. 2015). Hence, cooling needs to be seen from a wider vantage point, that is on a meso and indeed macro city level. There has been global interest for the idea of 'compact cities', due to advantages of mixed use and public transport optimisation (Burton 2000; Owens 1986). However, compactness can worsen urban ventilation and air pollution. A study of 45 core cities in China illustrated the impact of city compactness on various parameters such as infrastructure efficiency, public transport, residential energy consumption and environmental externalities (Chen et al. 2008). There are various effects of compact development; of special note in our context are two key aspects: (a) the impact on cooling energy consumption due to reduced ventilation and increased air pollution; and (b) the impact of reduced green and open spaces since these have significant cooling and pollution reduction functions (De Sousa Vale 2008; Akbari et al. 2001; Tratalos et al. 2007). To understand and optimise cooling strategies for urban development, seven potential socio-environmental problems of over-compact cities need consideration (Chen et al. 2008):

- Higher urban density implies less urban green or open space;
- Traffic congestion increases travel time, fuel consumption, UHI and poor air quality;
- Compact buildings normally have poor access to daylighting and natural ventilation cooling;
- High-rise residential blocks discourage community life and social communication with negative effects especially for the elderly and children;
- Overcrowding in compact neighbourhoods may result in escape to suburbs and decentralisation;
- Compact city buildings typically require energy intensive materials and technical services;
- High-density buildings may increase energy demand for lighting, ventilation and refrigeration.

In this study of clustered housing in India, we discuss compactness and green infrastructures in particular since these two parameters are keys to significant cooling and energy savings (Ignatius et al. 2015). This case study also highlights the importance of vernacular design and community setting.

Issues of Compactness

Compactness is a key question since hot-climate developing countries struggle with slums and informal settlements and many highly congested city environments, both low-rise and high-rise (Sanaieian et al. 2014). Many countries have informal settlements but they are widespread not least in the Asian region (Degert et al. 2016). Countries like India are particularly susceptible, having many highly clustered settlements lacking basic amenities (Roy et al. 2017). Studies have shown the effects of lack of access to water, sanitation and energy with poor health, thermal discomfort and high cooling needs (Ahmad and Puppim de Oliveira 2015; Degert et al. 2016; Parikh et al. 2012). How does the physical form of cities affect energy consumption and invariably emissions? On a community scale, urban form includes design elements such as housing size and types, density, orientation and building configuration, open space and landscaping (Jabareen 2006; Ko 2013; Næss 2001; Holden and Norland 2005). All these factors affect the microclimate including solar radiation, humidity, shade, wind flow, daylighting and local heat island effect (Ko 2013; Sanaieian et al. 2014).

In the 1970s and 1980s, factors such as, street orientation, density, vegetation and ground coverage became important topics in US urban planning (Holden and Norland 2005); and the term green infrastructure (GI) was introduced. GI regulates microclimate, both indoor and outdoor, by solar shading, modification of wind flows and the evapotranspiration process, and began to be applied in order to temper UHI (Killingsworth et al. 2011). A lack of GI is synonymous to increased temperature and air pollution (Hildebrandt and Sarkovich 1998; McPherson and Rowntree 1993; McPherson and Simpson 2003). As is also discussed in Chap. 7, trees in particular are noted to be cost effective with the added benefit of removing air pollution and improving air quality.

A dense spatial configuration is directly linked to heat gains and consequent high cooling demands, as well as poor air quality (Akbari et al. 2001; Johansson 2006). Compactness reduces urban air convection and increases heat absorption in cities, where air-conditioning (AC) often consumes about 40 % of building energy (Sailor 2006). In the inner cities, compactness means less open space and fewer or narrower streets or other channels for urban ventilation. Excessive compactness, thus, has a strong negative cooling impact, and there is a strong link between compactness and GI (Doick and Hutchings 2013; Golany 1996; Ignatius et al. 2015; Jabareen 2006). The goal of compactness has increased city density, sometimes backed by the environmental argument of reducing expansion into rural land; urban ventilation and cooling energy have been low priorities but are increasingly important (Dawodu and Cheshmehzangi 2017). Issues and consequences of city density are discussed further in Chaps. 8 and 13.

Clustered Housing and Blue-Green Infrastructures

Despite issues of pollution, ventilation and UHI, the design of compact settlements can still be greatly improved to create cooler cities. There is extensive research, examples of which are featured in this book, on shaping layouts and green infrastructures so as to provide the best response to natural conditions in order to provide healthy living environments, both indoors and outdoors, in our cities. Streets and pathways increase permeability, to assist passage of prevailing winds (Hughes et al. 2011); linking these and the open spaces as well as maintaining low-rise structures increases ventilation and reduces cooling requirements (Ng 2009). Street and building orientation play a major role; Ng (2009) suggests a staggering arrangement of building blocks. Other research suggests main orientation of streets within 30 degrees of prevailing wind conditions to maximise air flow. Building heights should be varied from high to low towards the winds (especially those prevailing during the hottest seasons) or else varying heights to limit stagnation and wind blockages (Ng 2009). Other factors include the geometry and air flow characteristics of street canyons (Shashua-Bar and Hoffman 2003) and the Height to Width ratio of streets; wide streets increase daylighting and solar radiation, which is needed for GI to thrive (Ko 2013; Norton et al. 2015). The city blocks and building volumes themselves should be as permeable as possible (Kamal 2012; Norton et al. 2015) with a maximum of openings near ground level, improving pedestrian thermal comfort and ventilating air pollution. Shading and reflective materials on roofs mitigate the intensity of solar radiation on buildings (Hughes et al. 2011; Okba 2005).

The use of GI provides shading and lower temperatures, especially effective being broad trees with dense canopies (Shashua-Bar and Hoffman 2003; Jim 2004). Increasing vegetation and GI is often problematic in hot-dry climates with water shortages (Gill et al. 2007), whilst others are affected by periodic droughts (Hughes et al. 2009; Srinivasan et al. 2012). Storage of storm water runoff for watering can provide a part solution, but is expensive (Coutts et al. 2013). Open spaces should where possible be linked together, to form wind corridors (Ignatius et al. 2015; Taleb 2014; Zhang et al. 2005). A study of Melbourne street canyon geometry (Matzarakis et al. 2010) found that east-west streets should be prioritised for trees. Narrow streets hamper the benefits of trees, hence, green walls and facades and ground level vegetation should be priorities. Not all plant species tolerate high gusts of wind and this occurs more in tall, narrow canyons (Norton et al. 2015).

Case Study: Malhar Eco-Village, Bangalore, India

Malhar Eco-Village is located in Kengeri, on the outskirts of Bangalore, south India. Bangalore has a fairly moderate hot climate due to its elevation but with intense solar radiation. The rapidly growing suburb is well connected with recent infrastructure in the area including hospitals, schools and other amenities which make it an attractive place to live. It has a good water table and abounds with farms and orchards. It is easily accessible, yet away from the pollution of the main roads. The project was designed and built by GoodEarth Company, a local practice focusing on sustainable communities in India. Their team of architects and builders endorse sustainable principles and construct environmentally friendly architecture. They have been experimenting with alternatives and concepts of holistic development integrating green-blue infrastructures, vernacular design and sustainable materials. Malhar eco-village promotes passive architecture and sustainable design by adapting to the site microclimate and topography, as well as blending conventional architectural design with traditions from the Bangalore tropical savannah climate (Fig. 7.1).

The project covers 20 hectares and is projected to accommodate families in 500 residential homes when completed. It forms a low-rise, quite dense pattern of clustered housing within a beautifully landscaped complex. It is divided into five smaller communities of between 60–90 houses to enhance human scale and manageability. Built in phases, some are already occupied and others are in progress. The five communities are described below. There are various house typologies including cluster homes, townhouses and apartments to suit different lifestyles and budgets. Each of the five communities is designed to be self-sufficient in its services and facilities, whilst other facilities are shared by all to enhance community interaction and to reduce dependency on the city. The buildings include green construction practices, vernacular techniques, natural materials and natural ventilation. The site development in all five areas displays integrated sustainable design including high biodiversity and use of indigenous species, rainwater harvesting, groundwater recharging and an eco-friendly, non-chemical sewage treatment plant. The building layout and orientation of all buildings is adapted to the topography so as to achieve an optimum cooling effect, thereby reducing energy consumption and providing comfortable conditions both indoors and outdoors (Fig. 7.2).

Clustered housing has had wide popularity in many regions, including Europe, for many reasons including economy and conviviality. It offers a moderate density and, often, common facilities and a good amount of vegetation and green space. Many clustered developments also have a common administrative structure, such as, housing associations or co-housing. In many cases, vehicle traffic is kept to the outside with garages and parking around the perimeter, creating car-free and noise-free spaces safe for play and recreation within the residential zone itself. In the following, we focus on three major aspects of the project. These are building form and layout as seen from the meso scale planning perspective; green and blue infrastructures; and materials and vernacular architectural design.

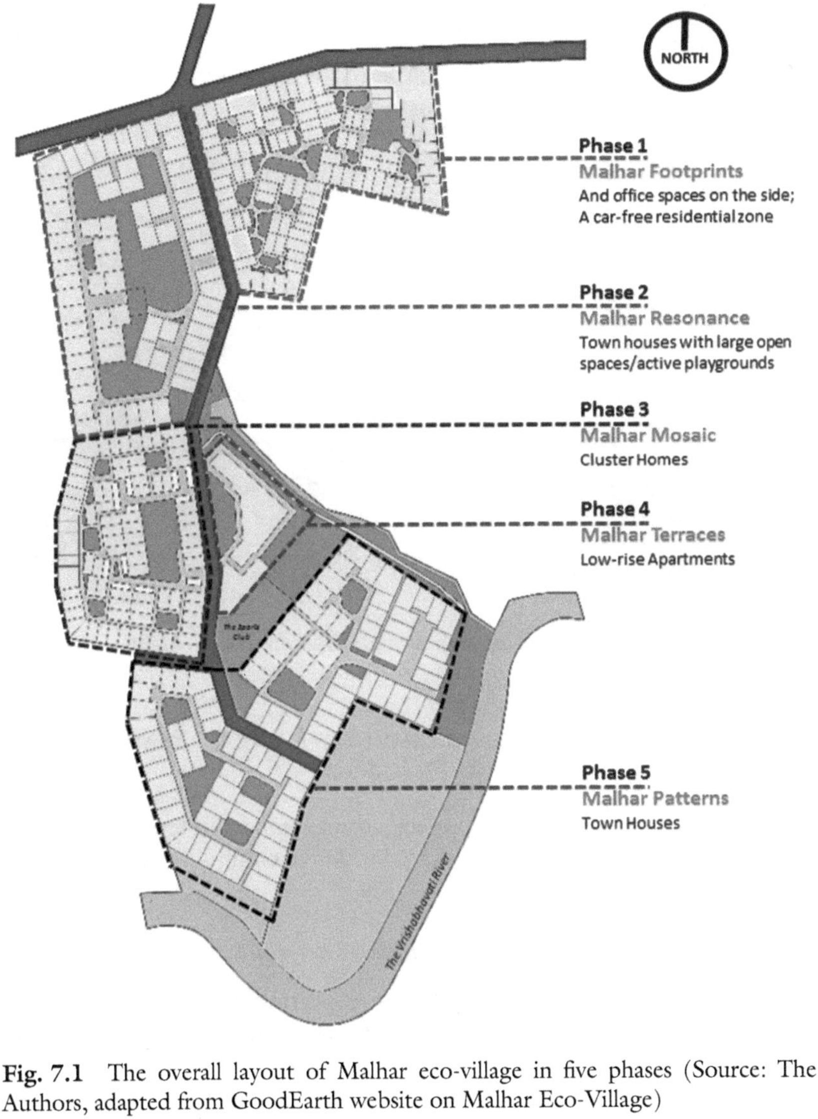

Fig. 7.1 The overall layout of Malhar eco-village in five phases (Source: The Authors, adapted from GoodEarth website on Malhar Eco-Village)

Fig. 7.2 The clustered layout of housing communities of Malhar eco-village (Source: The Authors)

Building Form and Community Layout (Planning and Spatial Layout)

To provide an overview of the project's achievements in terms of building form and community layout, we describe all five community areas.

Good Earth Malhar-Footprints

Completed in 2014, Good Earth Malhar-Footprints was the first phase, and has a cluster typology spreading over 3 hectares with 96 residential units of two- and three-bedroom houses. The design arrangement within a network of streets and landscaped courtyards provides a car-free environment. The cluster of spaces and interconnecting paths form a pedestrian-only space, safe for residents and for children. The cluster layout is based on groups of 9–16 houses around an open space like a large courtyard, with a shared parking area near one corner. 'The total open space in the project is 60 % of which the soft-scape comprises 48 % of the total site area' (Goodearth website). The sense of place is achieved through a hierarchy of spaces and variation of levels and the relationship between them, creating interesting transitions and avoiding monotony in the layout. The open spaces within the clusters are designed to encourage multi-use by the residents. The fairly dense layout provides shading for most of the walls, reducing heat absorption significantly. The pathways and openings between clusters draw breezes into and through the site. The need for artificial light and for air conditioning is eliminated for most of the year as the buildings are designed with adequate day lighting and natural ventilation.

Malhar-Resonance

Malhar-Resonance completed in 2015 has a townhouse typology spread over 4 hectares of land with 70 three- and four-bedroom units. The emphasis has been on street layout and interactive social spaces, with a pedestrian loop around the neighbourhood and a large playground. Adapting to the natural topography of the site, the spaces are on different levels. All units include a courtyard (either open or covered), split levels, atriums and verandas of different sizes. Early morning and winter sunlight are captured by layouts that utilise east and south facades. Similarly, west façades and entrance areas are designed to capture breezes from the west and to give good natural lighting for the main interior spaces.

Malhar-Mosaic

Malhar-Mosaic, currently nearing completion, has a cluster typology spreading over 2.3 hectares with 84 residential units consisting of three- and four-bedroom units. The layout emphasises a car-free approach, with a perimeter road that connects 15 clusters arranged around small parks. Here too, spaces are located at different levels with a diversity of courtyards, verandas, balconies and terraces. One cluster abuts a larger park that provides the community with a space for social interaction. In order to optimise light and natural ventilation, all units include either an internal or external courtyard, with an open floor plan facing gardens at both ends.

Malhar Terraces

Malhar Terraces has an apartments typology with 63 units on 1.2 hectares. Containing one-, two- and three-bedroom units, it is currently under construction. This is the first low-rise apartment type development in Malhar and accommodates different lifestyles and budgets. This area has a sloping site where the layout had to provide for resilience to flooding. The design incorporates receding terraces with small gardens staggered at different levels. Access to the blocks is by paths leading to the inner courtyards or atriums. These are richly planted. There are set-backs recesses in all units, to increase ventilation and lighting. The cross-ventilation is enhanced by the design of bay and French windows in bedrooms and living spaces, which are mostly oriented towards the gardens or balconies. The landscaped gardens form a terraced pattern to augment air movement between site levels and also act as a buffer against heat, dust and driving monsoon rain.

Malhar-Patterns

Malhar-Patterns is a townhouse typology spread over 5 hectares of land with 92 units consisting of three- and four-bedroom units, including four different types: open plan home, garden courtyard home, corner home and clustered home. This is also nearing completion. Most of the buildings are designed around open spaces with a street 'loop', again connecting the neighbourhood to provide interactive areas with limited vehicular traffic. The spatial pattern and scale of the units are varied through the size and orientation of buildings. There are rainwater collection and recharge zones, plantation areas and conservation zones for local and indigenous species. All houses are designed specifically in relation to the position and shape of the allocated plot. They also include internal and external courtyards with an open plan layout optimising ventilation and light. The courtyards or walled gardens are enclosed on two or three sides by windows and verandas. This area focuses on especially strong integration of greenery and trees. As in all five areas of the eco-village, the low-rise high-density approach enhances passive cooling and ventilation regardless of size and orientation of individual buildings.

Fig. 7.3 A typical water feature of Malhar eco-village housing community (Source: The Authors)

Blue-Green Infrastructures at Malhar

A very prominent feature of Malhar is its blue-green infrastructure, with local species and high biodiversity. These features are central in all five community plans and are accessible for various socialising functions. The natural features of the site are maintained and exploited to enhance both microclimate and the sense of community. Each community includes 'blue' features, either in the form of small ponds or areas of rainwater collection and water recycling; some of these are utilised for irrigation and flushing. In all zones, the landscaping and water bodies are designed for groundwater recharge and rainwater collection (Fig 7.3).

Most surfaces are permeable, with a series of natural swales and percolation pits designed for the soil to absorb water and allowing for more natural environments within the sites. Tank structures and services are under hard surfaces. The communities benefit from sub-surface aquifers that are utilised to maintain the water table and ground water. There are thus many areas of water collection and discharge, which also provide the housing with cooling effects from the water features in the area. Malhar also abounds with existing mature trees and new plantations, providing shading in the public areas and open spaces in particular; all housing units also have at least one tree providing specific shading.

Material Use and Vernacular Architectural Design

As noted, all housing units of Malhar eco-village are designed for adequate natural light and cross-ventilation. This includes key elements of passive cooling design such as large windows, wide verandas, air channels for circulation, skylights and *jaalis* (Indian latticed screens). Openings and windows near skylights and stairwells assist upward discharge of hot air by stack effect and increase the air flow through the internal spaces. In return, the skylights bring natural light, creating bright and airy spaces. All house plans are oriented with consideration to the solar angles and wind directions. The maximum use of soft surfaces and courtyards reduces heat island effects and ambient temperatures. Verandas with different scale and proportions also create transitional spaces between interior and exterior, whilst protecting interiors from direct solar radiation and heat; best locations for these being on west and south sides (Fig. 7.4).

Materials with low embodied energy were used as much as possible and energy-intensive, less sustainable materials such as cement and steel were minimised and replaced by earth, timber and stone. The main building material is compressed stabilised earth blocks for which the raw material is the soil from the site itself. Natural materials can be used in ways that require minimal processing energy and offer good durability, low maintenance, renewability, low cost and local availability requiring no transportation. There are no hazardous emissions. They are also easy to repair and have zero environmental impact after demolition. Compressed rammed earth walls as well as blocks are used in improved versions of a millennia-old tradition. The block production is on site with small machinery and organic stabilisers. The blocks are strong, durable and can be used in loadbearing construction. The base of many of the buildings is made of local stone. This selection of materials also aims to ensure that interior spaces are warm in winter and cool in summer.

Other materials include reinforced concrete for floor slabs, whilst stairs are built with steel, concrete and wood. In all housing units, a double or cavity roof is used to minimise solar heat gain. A lighter construction is used

Fig. 7.4 The front view (*top*) and side view (*bottom*) of clustered houses with mud blocks and shading from verandas and fast-growing trees (Source: The Authors)

for roofing of semi-open spaces where shading is required. Terracotta tiles are used for roofing, generally in a light orange-red colour with good thermal qualities; they absorb heat but quickly release it, providing cooler spaces below. Timber, being a sustainable material, replaces conventional high-energy materials such as aluminium and steel for doors and windows as well as some floors. The architects procured local timber from the surrounding community.

In general, great attention is given to local vernacular layouts and material use, which reduces the embodied energy of the construction. This vernacular architectural design not only respects and harmonises with the local culture, but adds modern environmental value in terms of low embodied energy, recyclability, better cooling performance, and healthy, comfortable living.

Other Passive Cooling Design Features

Site microclimate is influenced by land form, vegetation, water bodies, street widths and orientation, open spaces and built forms. Some of these are discussed in more detail in the following sub-sections.

Orientation

In vernacular architecture, orientation is a major factor in regard to solar radiation, heat gain and loss, daylight and natural ventilation. In Malhar eco-village, street widths and building orientation are designed to reduce direct solar radiation and maximise air flow in between and into the housing units. Clustered type buildings can also limit solar radiation by casting shadow on neighbouring buildings, as can be seen in many parts of Malhar. The street orientation is mainly parallel to wind direction, and locating the lowest buildings along wide streets also reduces barriers to air flow reaching those behind.

In a tropical climatic context such as Bangalore, north facades receive least solar radiation with minimum intensity, hence a north-south building orientation is preferable. This minimises east and west oriented façades which receive the most troublesome solar radiation in the hot summer season. On the other hand, the south façade receives welcome solar radiation in winter. In order to maximise natural ventilation, openings should face prevailing summer wind directions and be shaped to direct air flow indoors, such as with adjustable louvers, wing walls and other vernacular techniques. Inlet openings should be mainly situated on the windward side at low levels and outlet openings on the leeward side at a higher level. The overall placement and alignment of the buildings should allow wind flow around and in to all the buildings, as can be seen in the building layouts in all five areas of Malhar eco-village.

Building Envelope

There is a direct correlation between the thermal performances of a volume of space inside a building and the envelope enclosing it. This, the surface/volume ratio, is determined by the building form. Building form affects solar access and wind exposure as well as the overall heat loss and gain through the external envelope. A compact form receives less heat gain and has less heat loss. In regions like Bangalore, compact buildings with low surface/volume ratio are preferred, to reduce heat gains. The particular form also determines the flow of air around buildings, influencing the air flow indoors and thus also the cooling needs.

The buffering effect of heavy materials—high thermal mass—is exploited in Malhar to achieve thermal comfort indoors. The construction in earth blocks with their heat storing and heat conduction properties plays a major role. Earth construction is heavy, but can be beneficial in hot climates given good cross-ventilation and night cooling; in tropical climates it also has better humidity properties than materials such as concrete.

Roof

In all housing units, good roof overhangs, especially to the west, provide protection from solar exposure. Where natural light is needed, smaller overhangs are designed, mainly on north and south sides. These again are features of traditional architecture. Additional shading is provided by other elements, such as added arch frames or roof-mounted solar photovoltaic panels, which provide a second layer to the roofing, decreasing incident solar radiation on roof surfaces. The reduction in indoor temperature with the use of this double roofing system has been measured at around 4–5 °C. Similarly, stack ventilation is enhanced with vented skylights.

Evaporative Cooling

Evaporative cooling is used as a secondary cooling strategy to lower indoor air temperatures. The sensible heat of air is used to evaporate water thereby absorbing energy and providing cooling. Water bodies, ponds and fountain areas in the landscape of many of the outdoor and public areas all help reduce the air temperature around the buildings and reduce cooling needs.

Courtyards

In general, courtyards perform as heat sinks where temperatures in the courtyards are lower than the outdoor or street temperature. They are also a regional vernacular feature. Vegetation in the courtyards assists in shading and evaporation. The use of courtyards throughout Malhar helps to maintain air flows, as low-level cool air from the courtyards moves through openings into the internal spaces. This simple but effective design feature maximises daytime air flow into the units and inverts the process during the night hours.

Conclusions

In Malhar eco-village we see a passive vernacular design responding to the local climate of Bangalore and topography of the site; both of which underline the value of sustainable design approaches, not least to reduce the energy needs for cooling and provide a cooler living climate, both indoors and outdoors, at the meso level of neighbourhood planning. A series of features have been integrated into the design in order to maintain low indoor temperatures without relying on mechanical air conditioning systems. Site plan, buildings layout, landscaping, building design, construction materials and detail features all contribute to the same goal of sustainability. It is notable that many of the features described are modern applications of solutions, some of them ages old, from vernacular architecture in the region. And beyond being environmentally responsive passive habitats, the Malhar project also achieves a strong focus on sociable, harmonious community living. The approach that revitalises vernacular architectural design, local materials use and traditional construction methods, is much appreciated as a path towards more sustainable communities in the context of India.

Apart from the design approaches and cooling strategies discussed in this study, this clustered type of housing development represents an effective low-rise but quite compact typology with many sustainability advantages. Similar approaches are naturally just as relevant in inner city areas, but as discussed elsewhere in this book, are less easy to apply due to the microclimatic constraints of high urban density, high costs and the demands of high-rise construction. In the context of developing countries, the many millions currently moving to and living in extremely uncomfortable, hot conditions in informal settlements and slums will not, for a long time, be able to afford inner city high-rise type urban living places. In the hot regions of fast developing Asia, relatively compact low-rise towns and clustered neighbourhoods may offer some of the best solutions for low ecological impacts and costs, as well as the coolest urban environments.

References

Ahmad, S., & Puppim de Oliveira, J. A. (2015). Fuel switching in slum and non-slum households in urban India. *Journal of Cleaner Production, 94*, 130–136.

Akbari, H., Pomertantz, M., & Taha, H. (2001). Cool surfaces and shade trees to reduce energy use and improve air quality in urban areas. *Solar Energy, 70*(3), 295–310.

Barnes, D., & Foley, G. (2004). *Rural electrification in the developing world: A summary of lessons from successful programs joint UNIDO/World Bank Energy Sector Management Assistance Programme*. Available online via http://iis-db.stanford.edu/evnts/3961/Doug_Barnes_paper.pdf. Accessed 2 Mar 2017.

Burton, E. (2000). The compact city: Just or just compact? A preliminary analysis. *Urban Studies, 37*(11), 1969–2001.

Chen, H., Jia, B., & Lau, S. S. Y. (2008). Sustainable urban form for Chinese compact cities: Challenges of a rapid urbanized economy. *Habitat International, 32*, 28–40.

Coutts, A. M., Tapper, N. J., Beringer, J., Loughnan, M., & Demuzere, M. (2013). Watering our cities: The capacity for water sensitive urban design to support urban cooling and improve human thermal comfort in the Australian context. *Progress in Physical Geography, 37*(1), 2–28. https://doi.org/10.1177/0309133312461032

Dawodu, A., & Cheshmehzangi, A. (2017). Passive cooling energy systems SWOT analyses for energy-use reductions at three spatial levels. *Energy Procedia, 105*, 3411–3418.

De Sousa Vale, D. (2008). *Sustainable urban form, accessibility and travel: The relationship between polycentric urban development and commuting in Lisbon*. Newcastle: University of Newcastle upon Tyne.

Degert, I., Parikh, P., & Kabir, R. (2016). Sustainability assessment of a slum upgrading intervention in Bangladesh. *Cities, 56*, 63–73.

Doick, K., & Hutchings, T. (2013). *Air temperature regulation by urban trees and green infrastructure*. Research note document. Available at https://www.forestry.gov.uk/pdf/FCRN012.pdf/$file/FCRN012.pdf. Accessed 1 Apr 2017.

Fell, H.-J. (2012). *Global cooling : Strategies for climate change*. Boca Raton/London: CRC/Taylor & Francis [distributor].

Gill, S., Handley, J., Ennos, A., & Pauleit, S. (2007). Adapting cities for climate change: The role of the green infrastructure. *Built Environment, 33*(1), 115–133.

GNESD (Global Network of Energy for Sustainable Development). (2007). *Reaching the millennium development goals and beyond: Access to modern forms of energy as a prerequisite*. ISBN:978-87-550-3600-0. Available at http://www.gnesd.org/publications/other-gnesd-publications. Accessed 02 Feb 2017.

Golany, G. S. (1996). Urban design morphology and thermal performance. *Atmospheric Environment, 30*, 455–465.

Goodearth organisation webpage. Available at http://goodearth.org.in/projects/bangalore/. Accessed 4 Apr 2017 and 7 May 2017.

Hildebrandt, E. W., & Sarkovich, M. (1998). Assessing the cost-effectiveness of SMUD'S shade tree program. *Atmospheric Environment, 32*(1), 85–94.

Holden, E., & Norland, I. T. (2005). Three challenges for the compact city as a sustainable urban form: Household consumption of energy and transport in eight residential areas in the greater Oslo region. *Urban Studies, 42*(12), 2145–2166.

Hughes, B., Caulautit, K. J., & Shazhard, S. (2009). *Adapting buildings to meet the energy challenge*. Available at http://www.energy.leeds.ac.uk/uploads/documents/104_Hughes%20-%20paper.pdf. Accessed 4 Mar 2017.

Hughes, B. R., Chaudhry, H. N., & Ghani, S. A. (2011). A review of sustainable cooling technologies in buildings. *Renewable and Sustainable Energy Reviews, 15*, 3112–3120.

Ignatius, M., Wong, N. H., & Jusuf, S. K. (2015). Urban microclimate analysis with consideration of local ambient temperature, external heat gain, urban ventilation, and outdoor thermal comfort in the tropics. *Sustainable Cities and Society, 19*, 121–135.

Jabareen, Y. (2006). *Sustainable urban form: Their topologies models and concepts*. Cambridge, MA: Department of Urban Studies, Massachusetts Institute of Technology.

Jim, C. Y. (2004). Green-space preservation and allocation for sustainable greening of compact cities. *Cities, 21*, 311–320.

Johansson, E. (2006). Influence of urban geometry on outdoor thermal comfort in a hot dry climate: A study in Fez, Morocco. *Building and Environment, 41*, 1326–1338.

Kamal, M. A. (2012). An overview of passive cooling techniques in buildings: Design concepts and architectural interventions. *Acta Technica Napocensis: Civil Engineering and Architecture, 55*(1), 84–97.

Killingsworth, B., Lemay, L., & Peng, T. (2011). *The urban heat island effect and concrete role in mitigation*. Available at http://www.nrmca.org/members/ConcreteInFocus/Enviro%20Library/NRC-S0511_urban.pdf. Accessed 12 Mar 2017.

Ko, Y. (2013). Urban form and residential energy use: A review of design principles and research findings. *Journal of Planning Literature, 28*(4), 327–351.

Matzarakis, A., Rutz, F., & Mayer, H. (2010). Modelling radiation fluxes in simple and complex environments: Basics of the RayMan model. *International Journal of Biometeorology, 54*(2), 131–139.

McPherson, E. G., & Rowntree, R. A. (1993). Energy conservation potential of urban tree planting. *Journal of Arboriculture, 19*(6), 321–331.

McPherson, E. G., & Simpson, J. (2003). Potential energy savings in buildings by an urban tree planting programme in California. *Urban Forestry & Urban Greening, 2*(2), 73–86.

Næss, P. (2001). Urban planning and sustainable development. *European Planning Studies, 9*, 503–524.

National Sample Survey Organization. (2015, August 19). *Key indicators of urban slums in India. National sample survey 69th round*, July 2012–December 2012. New Delhi: Ministry of Statistics and Programme Implementation; 2013. Available from http://mospi.nic.in/Mospi_New/upload/KI_SLUM_report69round_24dec13.pdf [cited 2015 Aug 19].

Ng, E. (2009). Policies and technical guidelines for urban planning of high-density cities—Air ventilation assessment (AVA) of Hong Kong. *Building and Environment, 44*, 1478–1488.

Norton, B. A., Coutts, A. M., Livesley, S. J., Harris, R. J., Hunter, A. M., & Williams, N. S. G. (2015). Planning for cooler cities: A framework to prioritise green infrastructure to mitigate high temperatures in urban landscapes. *Landscape and Urban Planning, 134*, 127–138.

Okba, E. M. (2005). *Building envelope design as a passive cooling technique, at the international conference of passive and low energy cooling for the built environment*, Greece.

Owens, S. (1986). *Energy, planning and urban form*. London: Pion Limited.

Parikh, P., Chaturvedi, S., & George, G. (2012). Empowering change: The effects of energy provision on individual aspirations in slum communities. *Energy Policy, 50*, 477–485.

Phadke, A. A., Abhyankar, N., & Shah, N. (2014). Avoiding 100 new power plants by increasing efficiency of room air conditioners in India: Opportunities and challenges, June 2014. Available at http://eetd.lbl.gov/publications/avoiding-100-new-power-plants-by-incr; Clean Energy Ministerial, 'SEAD and India's

Bureau of Energy Efficiency Host Workshop on Enhancing Space Cooling Efficiency and Demand Response,' 6 Aug. 2014. Accessed 13 Apr 2017.

Roy, D., Lees, M. H., Pfeffer, K., & Sloot, P. M. A. (2017). Modelling the impact of household life cycle on slums in Bangalore. *Computers, Environment and Urban Systems, 64*, 275–287.

Sailor, D. (2006). *Mitigation of urban heat islands – Recent progress and future prospects.* Available at https://www.coolrooftoolkit.org/wp-content/uploads/2012/04/Mitigation-of-Urban-Heat-Islands.pdf. Accessed 04 Mar 2017.

Sanaieian, H., Tenpierik, M., Linden, K. V. D., Mehdizadeh Seraj, F., & Mofidi Shemrani, S. M. (2014). Review of the impact of urban block form on thermal performance, solar access and ventilation. *Renewable and Sustainable Energy Reviews, 38*, 551–560.

Shashua-Bar, L., & Hoffman, M. E. (2003). Geometry and orientation aspects in passive cooling of canyon streets with trees. *Energy and Buildings, 35*, 61–68.

Srinivasan, A., Ling, F. H., & Mori, H. (2012). *Climate smart development in Asia : Transition to low carbon and climate resilient economies.* London: Routledge.

Taleb, H. M. (2014). Using passive cooling strategies to improve thermal performance and reduce energy consumption of residential buildings in U.A.E. buildings. *Frontiers of Architectural Research, 3*, 154–165.

Tejero-González, A., Andrés-Chicote, M., García-Ibáñez, P., Velasco-Gómez, E., & Rey-Martínez, F. J. (2016). Assessing the applicability of passive cooling and heating techniques through climate factors: An overview. *Renewable and Sustainable Energy Reviews, 65*, 727–742.

Tratalos, J., Fuller, R. A., Warren, P. H., Davies, R. G., & Gaston, K. J. (2007). Urban form, biodiversity potential and ecosystem services. *Landscape and Urban Planning, 83*, 308–317.

Wang, N., Zhang, J., & Xia, X. (2013). Energy consumption of air conditioners at different temperature set points. *Energy and Buildings, 65*, 412–418.

Zhang, A., Gao, C., & Zhang, L. (2005). Numerical simulation of the wind field around different building arrangements. *Journal of Wind Engineering and Industrial Aerodynamics, 93*, 891–904.

CHAPTER 8

Integrated Urban Design: Study for a Housing District in Ningbo, China

Chris Butters and Ali Cheshmehzangi

Introduction

This chapter presents a concept design study for a large residential city block in Ningbo, south-eastern China. Developed by the authors, as part of the ELITH research program described elsewhere, it is a case study on the meso or intermediate level of urban planning. However, the approach was intentionally holistic, aiming for the much-to-be-desired integration of solutions on all three levels of micro, meso and macro. The concept design, therefore also includes, if in concept form only, questions of buildings design on the micro level, and the macro level of energy solutions; in this case, a district cooling system for the project. The goal of our study was primarily pragmatic; to show that even given current urban development models, with which we do not necessarily fully agree, more sustainable solutions are feasible. Current practice, in China as in many other developing countries, is driven by rapid economic growth and is often characterised by large-scale urban developments, often by private developers, with few or no environmental requirements and with a drive for maximum profits. These private

C. Butters (✉)
University of Warwick, Coventry, UK

A. Cheshmehzangi
The University of Nottingham Ningbo China, Ningbo, China

A. Cheshmehzangi, C. Butters (eds.), *Designing Cooler Cities*, Palgrave Series in Asia and Pacific Studies,
https://doi.org/10.1007/978-981-10-6638-2_8

market forces typically do not have much interest in, or time for, research and innovation. This study aims to suggest that even given this model of city development, very considerable improvements could be made without significantly reducing profits. The improvements we target are in the area of ecological site planning, climate responsive design and construction, energy efficiency and, overall, cooler city environments with improved health and comfort for citizens.

The main challenge is that of space cooling. The rapid spread of air conditioning (AC) represents the largest increase in energy use. Each individual AC unit ejects more heat into the city air, thus increasing needs for cooling as well as the urban heat island (UHI) effect, which is a major health and comfort issue. This study, thus, includes what is almost the only solution to combat UHI in hot climates: district cooling (DC), which in addition to being far more efficient can offer large savings in primary energy and climate emissions. The key is to address the problem of energy for cooling, not with individual solutions of small AC units but at the urban scale.

With well-designed bioclimatic site layouts and architecture, plus efficient lighting, cooking and appliances, one can greatly reduce energy needs in buildings. Adding renewable energy in the form of building-integrated solar photovoltaics can further reduce the energy needs. Selecting simpler types of construction and materials lowers the embodied energy as well. This study does not illustrate energy efficient and bioclimatic design in detail but focuses on integrated solutions at urban design level. The starting point is that of typical large, high-rise residential city blocks. Whilst we question that approach, lifetime energy and carbon footprint of typical developments can be reduced by over half, using not advanced or experimental solutions but recognised green principles.

Trends in City Development

The energy and climate impacts of urban development in the fast expanding cities of countries such as China is a concern. In many countries, urban development is not subject to energy efficiency or other environmental requirements and is being planned and constructed following outdated and energy-intensive models. This implies a huge energy and emission burden for such cities in the future. In addition, many housing developments pose problems of quality including fragmentation, gated communities and social segregation.

Many new city areas are planned with little consideration as to improving urban microclimate, maximising local breezes or enhancing outdoor conditions with the help of the layout, landscaping and vegetation. The high-rise buildings themselves are often poorly oriented in regard to sun and wind and have one-sided apartments with inadequate daylighting and little or no cross-ventilation. In addition, most high-rise structures are very carbon intensive (i.e. mainly concrete, steel and glass) and lack energy efficiency measures. Some greener options are directly cost saving, and others are inexpensive. These are missed opportunities where the only major requirement is better awareness and planning amongst investors, property developers and designers. Design guidelines and binding building codes are, where feasible, very important drivers of change too. As is noted elsewhere in this book, however, in many developing countries such planning guides and regulations are not in place and there are not as yet the resources or governance mechanisms to implement them.

We discuss issues of urban form and density further in Chap. 12. Whilst we do not address the issue of urban transports here, it is well known that the overall urban layout and location of work/residential/service zones also has a major impact on energy use in cities. Key principles as to mixed use zoning and sustainable mobility are well established in recent ecocity experience, although difficult to implement as long as the private car remains such a high priority. Our focus in this proposal has been of a local nature, that is, to minimise vehicle traffic within the site itself, creating less polluted, quieter and safer spaces as well as reducing local heat emissions.

Choices of City Development

It is well known, and lamented, that principles for good, sustainable urban planning are seldom applied even though they have been recognised for a long time. The barriers are many and range from financing and land policies to conservative thinking or adoption of inappropriate paradigms—such as, the ideals of the skyscraper city and equally that of the suburban individual bungalow. There are basic questions to ask about what kind of city we choose. Chinese and other developing cities often aim for a very high density. But as many studies show, dense high-rise does not, in fact, necessarily house more people, and has some considerable disadvantages especially in sustainability terms. This is discussed and referenced further in Chap. 12. In this chapter, we refer to Floor Area Ratio (FAR) and Surface Coverage (SC), two of the most common measures of density. FAR, also often termed plot ratio, expresses the total amount of floor area on a given site, for example a FAR of 1.0 means 100 square metres of floor space per 100 square metres of plot. This could be a two-storey building with 50 square metres on each floor on a site measuring 100 square metres; and thus, occupying half of the site, hence the SC would be 0.5, which means the building footprint is half of the total plot area. Following an analysis of various urban typologies and of cities elsewhere, we posit here an overall density or floor area ratio (FAR) for our case site of around 2.5. This is somewhat lower than some recent Chinese high-rise neighbourhoods, but we consider that as a general rule, residential densities above FAR 3.0 can imply a range of environmental as well as social disadvantages. It is notable, also, that traditional low-rise European cities often achieve this density or even higher—up to FAR 4.0 in parts of Paris for example. There is no inherent need to go high-rise in order to house large populations. We have, therefore, chosen to illustrate an urban block layout that has elements of high-rise as well as low-rise solutions. Whilst not ideal, this offers useful comparisons, and illustrates significant climatic and energetic—as well as economic and social—advantages of low-rise options.

High-rise is not cheap. It necessitates complex building structures as well as infrastructures, which are particularly expensive in relation to factors, such as, fire safety, ventilation, façade maintenance and, ultimately, demolition. Many high-rise residential blocks are virtually gated communities (Fig. 8.1)

Fig. 8.1 Typical high-rise residential typology in the city of Ningbo, China (Source: The Authors)

and even where they have a few common facilities, these are often underused.

As is often argued, high-rise does enable large areas at ground level to be left open; but it can be noted that this additional green space is not a 'natural' environment; it is usually only a thin 'green' layer on top of large concrete underground parking garages. These landscaped areas can be attractive but of limited functionality; they often appear to be designed for aesthetics rather than to create meeting spaces or improve the outdoor microclimate. By contrast, the classic European city model has equal or even higher population densities, but green open spaces are provided mainly in the form of large (and fully public) urban parks, within a short distance, which provide more functions and social meeting places as well as being genuinely green and large enough to provide a cooler city microclimate at the meso level.

Pragmatic Considerations and Relevance Elsewhere

The present study illustrates a proposal for a large city block in Ningbo, China. Whilst our key focus is hot-climate cities, Ningbo is hot and requires cooling for several months of the year but has some winter heating need too. The principles embodied in our study have even more relevance in hotter climatic contexts, such as many tropical Asian cities where cooling is needed all year round. It may be remembered that even many temperate climate cities, in Europe and elsewhere, also now have large cooling needs in summer, and UHI in those climates is also leading to increasing issues of public health including discomfort and indeed rising mortality during extreme heat events. This is partly due to inappropriate design, such as, heavily glazed buildings, but also due to rising comfort demands and behavioural changes. It is further exacerbated by a warming global climate; one of the widely documented features of which seems to be increasing frequency of extreme weather events. Similarly, the environmentally conscious layout and design principles illustrated in our study are broadly relevant to urban planning in other contexts. On the macro level of city energy solutions, the district cooling system shown is in the Ningbo case based on river water, but such DC systems can be even more advantageous in tropical locations; especially, with a large heat sink such as near the ocean. Large DC systems are found in several Asian cities.

Description

On a riverside site of about 15 hectares in Ningbo, presently vacant—but subject to near future development—we illustrate a layout comprising around 370,000 sq. metres of residential building, containing some 4600 apartments, plus common facilities and, as is a widespread custom, some commercial premises in low 'podiums' along the street perimeter of the site. This equates overall to a FAR density of 2.5. There are extensive areas of green space between the buildings and along the river bank which forms a curve around the top end of the site. The layout is designed to maximise prevailing local breezes, as well as to ensure access to sun in winter and solar protection in the hot seasons. Developing the detailed site layout and design will, however, always depend on more specific local studies and simulations of seasonal insolation, wind and air movement, ways to maximise site ventilation, selection of appropriate vegetation, design of permeable and reflective surfaces, optimising stack effect, noise reduction and other parameters.

About half of the apartments are in fairly high-rise towers; in contrast to common practice, these are designed so that nearly all apartments have light and ventilation from at least two sides. There are almost no units with exclusively north-facing exposure, which commonly fetch considerably lower prices. It is notable that many current high-rise developments do not even adopt such basic measures that would not only improve the living qualities and energy performance but offer higher profits. There are no one-sided apartments which are facing only west or east either, orientations that also cause unwanted overheating in respectively the evenings or the mornings. Overhangs and balconies provide summer shading as well as enhancing air flow; balconies can help to 'capture' more of the incident wind and lead it into the buildings. Most of these are oriented to capture prevailing summer breezes. Due to the provision of lighting and ventilation from two sides, the apartments do not need the extensive glazing that one often sees, needed to achieve daylighting in the very common one-sided narrow apartments. In many current blocks one sees that the entire front wall is of glass, inevitably causing overheating problems in summer.

The high-rises illustrated on the site are of up to 30 floors; they are staggered with increasing heights towards the north end of the site, this pattern of heights being to favour access to prevailing breezes from the south. The high-rise part of the site requires large underground parking areas, which extend far out under the site beyond the buildings themselves.

Fig. 8.2 The aerial view of our proposed design for a large residential city block in Ningbo, China (Source: The Authors)

This is in contrast, commented on below, to the case of the low-rise typology on the other part of the site. Facilities, such as, common rooms, gyms and so on, are located on the bottom floors of these towers, and a kindergarten is provided near the centre of the site with access to playground, gardens and riverside (Fig. 8.2).

The low-rise area, with apartment buildings, of 4–6 floors, illustrates a simple and cost-effective building typology. Here, parking is only partly below ground level, allowing through ventilation cooling the undersides of the apartments. The parking is also very inexpensive since most of it forms the necessary foundations and base walls of the buildings. Given this typology, the parking area needed corresponds to no more than the footprint of the apartment buildings themselves. In this low-rise solution, ground floor apartments are raised around one metre above ground level, providing privacy to the ground floor apartments and their balconies, but equally to enable cross-ventilation through the garages (Fig. 8.3).

We underline that that this is a concept layout only, the main focus being on the meso level of site layout not architectural solutions. The relatively similar rows of low-rise buildings illustrated here would, at the detail design stage, be provided with more interest and variation both in the way they are grouped and in design detailing. However, the important underlying

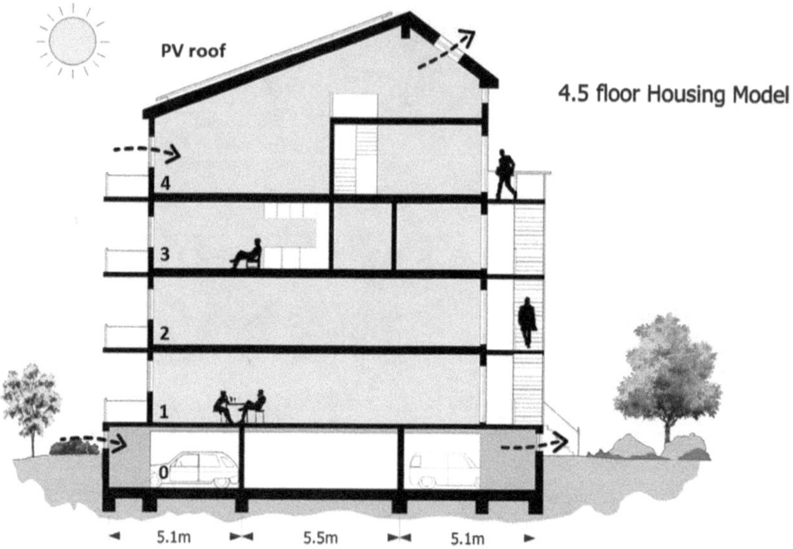

Fig. 8.3 Section through a typical low-rise typology of 4–6 floors (Source: The Authors)

principle is that all apartments have access to sun, light and prevailing breezes, and are spaced so that they do not impede sun and air movement to those behind. Note that in this low-rise layout, there are almost no façades that receive troublesome east or west sun. Similarly, such low-rise buildings can be shaded (for free) by moderately tall trees, whereas this is impossible for high-rises. Crucially, all apartments are cross-ventilated. This enables natural ventilation, whereas in one-sided apartments, mechanical cooling is an unavoidable necessity.

The low-rise typology provides apartments of various sizes and flexibility for different tenants. The apartments are accessed by simple (but covered) stairways. For the elderly and wheelchair users, access with lifts would be provided in a small percentage of the buildings. The spaces between the low-rise rows form semi-private 'courtyards' with social functions, not unlike the tradition of small courtyards in Chinese villages, which are much used outdoor spaces for gardens, childrens' play, relaxation and meeting (Fig. 8.4).

Studies in detail have been made of which (indigenous) species of trees and shrubs are most effective for filtering air pollution. Another research

Fig. 8.4 Perspective of low-rise courtyard and outdoor spaces for 4–6 floor housing units (Source: The Authors)

study of this type is described in Chap. 7. The role of the low commercial podiums along the street edge is an important topic, which is also addressed in the research study in Chap. 10. These podiums are in general unfavourable for cooling, since they form a barrier to air movement at street or pedestrian level. At the stage of detail design, therefore, they need to be designed carefully and with frequent openings. For example, such passages through to the site can be shaped like 'funnels'. Seen from the street side, this provides increased shop window areas and at the same time focuses or channels more air in through the passages. On the other hand, if a street is very polluted it may not be desirable at all to funnel street air in to the site. And whereas podiums may be unfavourable for air movement, they may be useful for their barrier function as regards noise. This illustrates the complexity of urban design issues.

Energy and Carbon

The buildings themselves are designed using passive strategies whereby the building design and construction leads to much reduced cooling needs. This includes maximum cross-ventilation, a lightweight reflective 'sun-screen' façade cladding, some lightweight materials, effective shading and avoiding excessive glazing, and insulation on external walls to reduce diurnal heat gain. In the high-rises, the stairwells are, further, designed to enhance stack effect ventilation and contribute to passive cooling of the buildings. Having done this as the first order design priority, the next step is to assist cooling by adding the simple technology of ceiling fans, which consume very little energy and can further provide a cooling effect of about 3 °C. The final, remaining need for space cooling is then provided by cool water circulated to the apartments from the district cooling plant, situated near the bend in the river.

Undesirable internal heat gain in the apartments are also minimised. This is achieved partly by selecting the most energy-efficient electrical appliances and LED lighting, plus ensuring better natural daylighting than in typical high-rises. The heat from cooking is a major source of indoor heat. In addition, comes the energy consumption of kitchen air extractors for the cooking heat as well as cooking odours and grease. In the low-rise blocks, provision is made for balconies where some semi-outdoor cooking can be done during the hot seasons. Sustainable solutions seek to identify and build on local opportunities; features such as cooking practices are typically culture-dependent. In countries, such as, Thailand it is quite common; in other countries, less so.

Given improved daylighting as well as lighting and electrical appliances of high energy efficiency, average household demand can be as low as 30–40 kWh per square metre of apartment. For a low-rise typology such as this, photovoltaic (PV) solar systems on the south-facing roofs provide a large part of the remaining low energy requirements for electricity, around 1500 kWh per apartment. Solar roofs on high-rises by contrast would only provide a small fraction of the energy needs.

Regarding embodied energy (or carbon), it has been noted earlier that this is an increasing part of the total energy/climate impact of buildings. Choices of building materials can reduce this part of the resource footprint of buildings. But it is easier to reduce this impact in low-rise structures, where one can use construction materials with considerably less embodied energy/carbon. In high-rises, it is difficult to avoid reinforced concrete and steel; and these two materials alone often constitute well over 75 % of the total embodied carbon in large buildings. Even so, the embodied footprint of the high-rises is also significantly reduced in this proposal compared to current models. Note also that high-rise necessitates large areas of underground parking and these structures, and the thick concrete decks above them, are both very expensive and extremely carbon intensive. This little noticed issue—the carbon footprint of site infrastructures—has been explored by us elsewhere (see Chap. 12).

Other interventions can include planted 'green roofs' on the north slope of roofs, reducing unwanted heat gain; these together with permeable paving surfaces at ground level also reduce the load on city drainage systems during heavy rains, and are indeed becoming mandatory in some cities. Water bodies and ponds also ameliorate the microclimate. Hence, a range of design features reduce energy needs and contribute to a comfortable microclimate in such areas of the city. The entire site design of landscaping, vegetation, water features and overall layout maximises cooling, enhances site ventilation and provides a minimum of hard paved surfaces which would increase local outdoor temperatures. The design also has a minimum of internal roads; and no surface parking except for services and deliveries, plus some parking along the main street shopping zone—outside the residential zone itself. The energy plant with the river-based cooling system is located on the bend in the river.

District Cooling

As described in more detail in Chap. 11, district energy systems can be very favourable. District heating (DH) systems are common now in temperate climates: in some cities there are district energy systems that provide both summer cooling and winter heating. However, district cooling (DC) systems are as yet not at all widespread. The source of cooling may be sea, river or lake water or the underground; they use standard heat exchangers, pumping and distribution technologies. It should be noted that given good planning, rejecting the waste heat to the sea or a large river normally has quite small ecological impacts at this scale.

With regard to energy and climate emissions, DC has four major advantages: (a) the larger scale offers efficiencies double that of small AC units; (b) the cooling source, such as, river water, is at a lower temperature than the ambient air in the hot season, and this difference ('delta-T') again provides an energy saving; (3) these systems offer much superior management and control than individual systems; (4), and not least, the waste heat is rejected outside the city itself, in this case via the river, so that there is no heat island effect as there is with thousands of individual AC units; on the contrary, the system contributes to cooling the whole local environment. This is a sustainable solution at the macro level. This unique feature of actually reducing UHI is why district cooling is gaining much attention now, and offers a central pillar of energy policy for hot climate cities.

Conclusions

The expanding cities of hot-climate developing countries often offer uncomfortably hot environments, and are the fastest growing source of energy consumption and climate emissions. In addition to poor outdoor conditions, much of what is being built today is of an outdated and inefficient standard. This represents a great missed opportunity; cities are locking themselves into huge future energy and climate burdens in addition to poor public health. Whilst only illustrated briefly here, the study suggests that these can be avoided by better planning; and that this does not require new knowledge, added costs, or advanced technical solutions.

High-rise city ideas are popular. This may be desirable for central business districts, but for other districts and residential areas there are a number of potential disadvantages, both social, economic and environmental. Further, one may achieve equally high population densities with low-rise typologies. It is not difficult to plan and build more sustainable housing. The potential energetic and climatic savings to the cities, to consumers and to society as a whole are very large indeed. Given growing political commitment to energy and emission reductions, it is urgent that solutions of this kind be better understood and widely applied. Cooler cities in hot climates are possible.

Reference

A broad selection of references is provided in other sections of this book on the topics discussed in this chapter. We also refer the reader to the ELITH web site. http://www2.warwick.ac.uk/fac/sci/eng/elith/ where several of our research publications provide extensive sourcing.

PART III

Case Study Chapters: Macro Level

This final section addresses the macro scale of entire city districts. First we present a research study from Singapore into city block structures and building patterns in order to maximise urban ventilation in hot cities. Then we provide an example of concrete planning measures that are now being applied in several cities to coordinate green spaces, transport corridors and other open avenues, as well as built volumes in order to ameliorate the urban climate. The final chapter in this section gives an overview of the important macro scale solution of district cooling systems for large urban areas, with Malaysia as particular example.

CHAPTER 9

Modeling City Patterns for Urban Ventilation: Strategies in High Density Areas of Singapore

Abel Tablada and Yueyang He

Introduction

By 2050, 70 % of the world population is expected to live in cities, nearly half of these in the tropics (United Nations 2017). The current population growth in tandem with the demographic transition from rural to urban has two consequences: on one hand, a continuous expansion of existing urban areas creating low-density suburbs of unequal quality of planning, infrastructure and services; and on the other hand, the urban core areas suffer a gradual process of land intensification, with increasing population and building density. This process is particularly intense in several emerging tropical and subtropical Asian cities.

For this huge number of tropical urban dwellers, one of the most significant threats is thermal discomfort, caused by the relatively high temperature and humidity as well as the exacerbation of warmer conditions from the urban heat island (UHI) phenomenon (Wong and Chen 2008). Further rising of average and extreme temperatures is expected due to climate change. In the case of Singapore, the increase is expected to be between 1 °C and 2 °C by 2050 compared to the period of 1986–2005

A. Tablada (✉) • Y. He
Department of Architecture, School of Design and Environment, National University of Singapore, Singapore, Singapore

A. Cheshmehzangi, C. Butters (eds.), *Designing Cooler Cities*, Palgrave Series in Asia and Pacific Studies,
https://doi.org/10.1007/978-981-10-6638-2_9

(Stocker et al. 2014). However, even with more pessimistic scenarios, urban planners and architects should still consider passive cooling means to reduce as much as possible the use of air conditioners, to minimise the current cycle of higher UHI effects, higher temperature, higher greenhouse gases (GHG) emissions and more acute climate-change impacts. Fortunately, several methods are proven to reduce UHI effects, to cool physical surfaces and improve thermal comfort. Amongst them, natural ventilation has been the most common cooling method in the tropics and is thought to remain one with high efficiency under certain conditions, and with zero energy consumption.

Singapore has a typical rainforest tropical climate, with mean daily maximum temperature and relative humidity ranging from 30 °C to 32 °C and 94 % to 98 % in the past 80 years (National Environment Agency 2016). Under such conditions, wind is crucial to improve thermal comfort from both physiological and psychological points of view. According to field survey results in Singapore (Yang et al. 2013), over 70 % of respondents felt thermal discomfort with low wind speed even though they were in outdoor shaded areas. In contrast, according to another study by Givoni et al. (2006), with higher wind speed of 1 m/s, people can tolerate higher outdoor temperature of up to 32 °C in the tropics. According to previous studies (Hong Kong Planning Department 2005; Building and Construction Authority 2010; Yang et al. 2013), a minimum wind speed of 0.8–1.5 m/s is required for achieving thermal comfort in shaded outdoor environments. Besides, as Chatelet et al. suggested (1998), indoors should be at least as comfortable as the outdoor climate when no mechanical cooling is used. According to BCA (Building and Construction Authority 2012), a minimum wind speed of 0.6–1.0 m/s is required for good indoor natural ventilation in Singapore.

Although wind speeds in unobstructed areas in Singapore are relatively low (yearly-averaged 2.7 m/s at 15 m height for prevailing wind directions), they are high enough to satisfy thermal comfort in most periods. However, large-scale constructions and the extension of urban areas have a substantial impact on the wind flow patterns, not only obstructing the cooler air from entering the city, but also stagnating the warmer air in the urban canyons. The most affected area in Singapore is the central business district (CBD), where a maximum temperature difference of 4.01 °C from the non-building areas was recorded (Wong and Chen 2005).

To make good use of natural ventilation, buildings and breezeways should be arranged with careful strategies in high-density urban areas.

Existing strategies mainly focus on: (a) building configuration; (b) street canyon geometry; (c) building disposition; and (d) urban density.

Building configuration focuses on a small spatial scale and aims to improve the indoor and immediate outdoor environment by adjusting and refining the building design. However, some of the proposed strategies have an impact at a larger scale. Building orientation according to prevailing winds, building form and indoor layout are the most common design guidelines in building codes to ensure minimum requirements for cross ventilation. For high density and high-rise morphologies, Ng (2009) recommended increasing the wind permeability by avoiding large scale podiums and introducing air passages and setbacks inside a plot. Regarding building forms, Li et al. (2013) found that rectangular buildings perform better than H-shape buildings because of their higher resistance to the surrounding airflow generated by complicated building shapes. However, H-shape buildings may provide enough façade area to enhance cross ventilation.

Street canyons also belong to the micro scale level but their different geometrical features may also have an impact at the meso scale level. The study of street canyons aims to optimise airflow by adjusting their height, width and length for better ventilation and pollutants' dispersion (Cheshmehzangi et al. 2010). After analysing the urban morphometry of some global cities, Oke (2002) found that the airflow pattern inside a street canyon is actually related to its aspect ratios, namely the ratio of height to width (H/W) and the ratio of length to width (L/W). Later, a more complete mathematical model (Georgakis and Santamouris 2005) was developed to predict the wind speed inside a street canyon by considering the geometries of the canyon, the wind speed and the wind direction outside the canyon.

At a medium to larger spatial scale, building disposition strategies aim to benefit natural ventilation by adjusting the relative position of buildings in a cluster or urban plot. Tsutsumi et al. (1992) showed that a staggered building array is worse than an aligned building array for generating building ventilation. However, contrary to their results, Zaki et al. (2012) suggested that the staggered patterns have better building ventilation potential than aligned patterns. These contradictory results are probably due to the different building intervals and aspect ratios that lead to different sheltering effects on the buildings downwind. Rajagopalan et al. (2014) suggested that step-up building configurations distribute wind evenly and allow the wind to reach the leeward side of each building on top of the podium.

The final focus uses building density indicators to assess urban wind permeability and the potential for natural ventilation at the urban scale. Kubota et al. (2008) found that lower building densities tend to ensure higher wind speed at the pedestrian level especially for neighbourhoods of detached houses. Buccolieri et al. (2010) indicated that, in urban areas with higher densities, more reverse flow and stronger vertical recirculation were found inside deep street canyons. As a result, lower horizontal flow rates and larger local mean age of air at ground level were found. At a large spatial scale, building densities can be seen as the urban surface roughness and used to detect air paths by identifying the high-density and low-density areas in a city or region (Wong et al. 2010).

Although the abovementioned strategies and foci are relevant for planners and urban designers as a guide to achieving effective urban ventilation, they often fail to describe the most influential features of the urban wind behaviour. Therefore, in this chapter, new strategies related to the breezeway network patterns are assessed in terms of their contribution to the cooling effect of urban ventilation in high-density areas. In contrast to the earlier-mentioned strategies, those proposed here focus on the structural features of breezeway networks. This is highly relevant taking into account that road network patterns and morphological features are primary considerations in the urban planning process and are crucial for the urban wind behaviour.

Nevertheless, for effective cooling by urban ventilation, it is not only the urban morphology that has to be adjusted. A necessary shift for the success of passive cooling at the urban scale should include the reduction of threats to natural ventilation in contemporary urban areas. Related problems like air pollution and noise from motorised transport should be minimised by implementing measures that favour walkability and clean transport modes.

Therefore, in a scenario of resilient buildings and neighbourhoods in which mixed-mode ventilation is applied in residential, commercial and office buildings, the assessment of urban ventilation and its cooling potential should include the prediction of both outdoor ventilation at pedestrian level and indoor ventilation at upper levels. Focusing on one while sacrificing the other target could minimise the cooling effect of urban ventilation in dense urban environments as well as lessening the effects of carbon footprint reduction measures in tropical cities.

New Strategies and Methods

Strategies of Adjusting Breezeway Network Patterns

Singapore's CBD is one of the most high-density urban areas in the tropics. The selected built-up area (Fig. 9.1) is occupied by compact and large building bulks with a very high gross plot ratio (GPR) of 12. The breezeways are arranged in an orthogonal pattern with shorter breezeway spacing aligned on a southeast-northwest axis.

Five strategies aiming to improve ventilation conditions at pedestrian and upper building levels in urban areas with the same GPR were tested using computational fluid dynamic (CFD) simulations. These five strategies are: (1) reducing breezeway network densities, (2) varying breezeway widths, (3) varying breezeway spacing, (4) staggering the blocks and (5) inserting T-shaped junctions by merging two plots.

Urban Models and Computational Settings

Twelve urban models were created to assess the five strategies. As shown in Fig. 9.1, two baseline models (A0 and B0) were first developed according to the breezeway network pattern in the selected area. Their plots are occupied by either separate blocks on top of the podium (type A) or separate blocks on the ground (type B) with 40 % and 20 % building coverage ratio (BCR), respectively. Subsequently, five other models were developed from each baseline model (A1–A5 and B1–B5) by adjusting their breezeway network patterns with the abovementioned five strategies. Meanwhile, the GPR and BCR are kept constant.

Regarding the computational settings, the approached wind at different elevation heights are estimated by a reference wind speed of 2.7 m/s at 15 m height. This value corresponds to the yearly-averaged data from Changi airport in Singapore. The wind profiles were estimated by using the logarithmic law equation with roughness length of 1 m. The four dominant wind directions in Singapore, recommended to be considered by BCA in CFD simulations, are north (N), northeast (NE), southeast (SE) and south (S) (Building and Construction Authority 2012). Turbulence is estimated by the commonly used standard κ-ε model (Launder and Spalding 1974). The computational domain and settings follow Franke et al. (2007), Architectural Institute of Japan (AIJ) (Tominaga et al. 2008) and Blocken et al. (2016) recommendations for urban ventilation analysis.

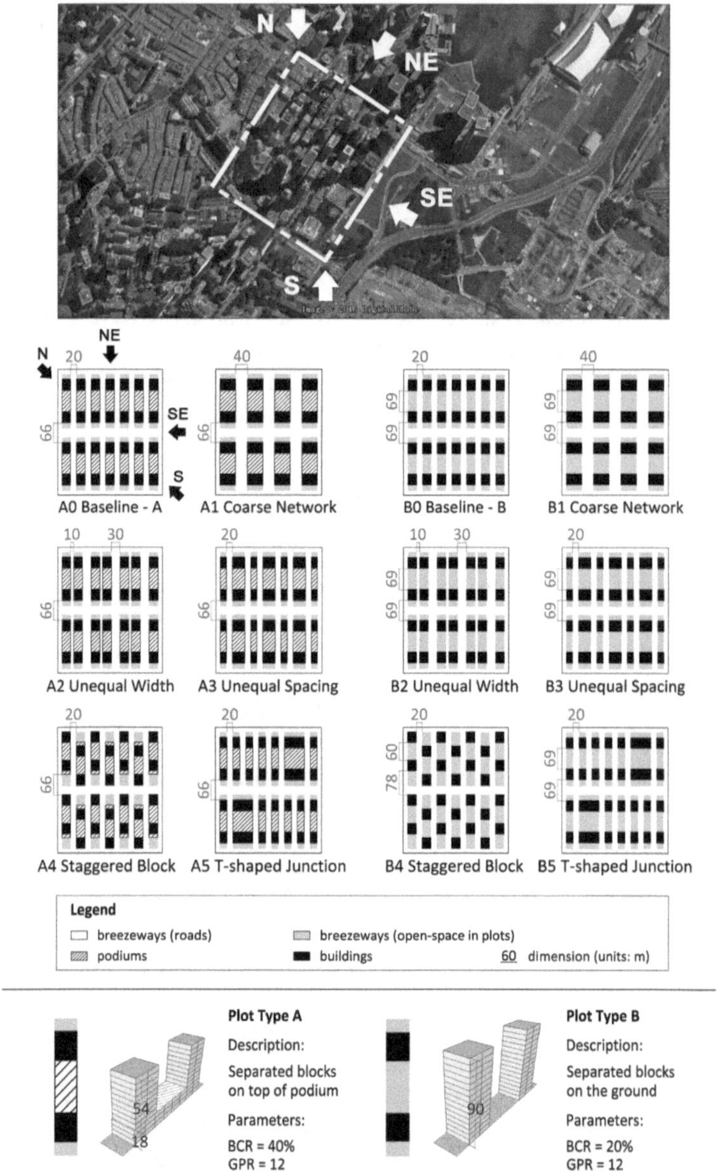

Fig. 9.1 Urban models for two types of plots and five strategies related to breezeway network patterns (*BCR* building coverage ratio, *GPR* gross plot ratio)

Evaluations of Natural Ventilation

To comprehensively evaluate the urban ventilation conditions, this study obtains wind data at both pedestrian and upper building levels. It is assumed that the best urban ventilation should benefit both outdoor and indoor activities.

Wind speed is commonly used to evaluate ventilation efficiency by urban ventilation assessment systems as well as green building standards and certification systems. It is easily understood by planners and architects and more importantly, it is directly related to human thermal comfort. Therefore, this study assesses outdoor wind conditions at pedestrian level (2 m height from the ground). Specifically, it calculates the average pedestrian wind velocity (U) at the selected area of each urban model. For indoor wind conditions, however, the estimation of wind speed to assess the ventilation potential is not sufficient since wind speed is affected by both outdoor and, most of the time, unknown indoor environments. Therefore, this study uses the difference of wind pressure on building facades to indicate the indoor ventilation potential. Specifically, it calculates the average wind pressure difference (P_d) of all buildings above podium level in the selected area. P_d is calculated by obtaining the pressure (Pa) at opposite facades, where operable windows are assumed to be installed (Fig. 9.2).

Effects of Five Strategies

Coarse Breezeway Network

The first strategy reduces the number of breezeways parallel to NE wind direction from 8 to 5. At the same time, the breezeway widths increase from 20 m to 40 m.

For pedestrian ventilation, compared with the baseline networks (A0 and B0), the values of U obtained on cases A1 and B1 increase when wind comes from an oblique direction (N and S); and they decrease when wind is from SE. This trend is more obvious for the case with no podium structure (B1). For building ventilation potentials, when wind is from SE, N and S, evident increments of P_d are seen with decreasing breezeway network density. However, for NE, where openings on facades are parallel to the wind direction, P_d remains extremely low as for the baseline cases.

The results from this strategy clearly suggest that, although keeping BCR constant, introducing a breezeway network with fewer but wider

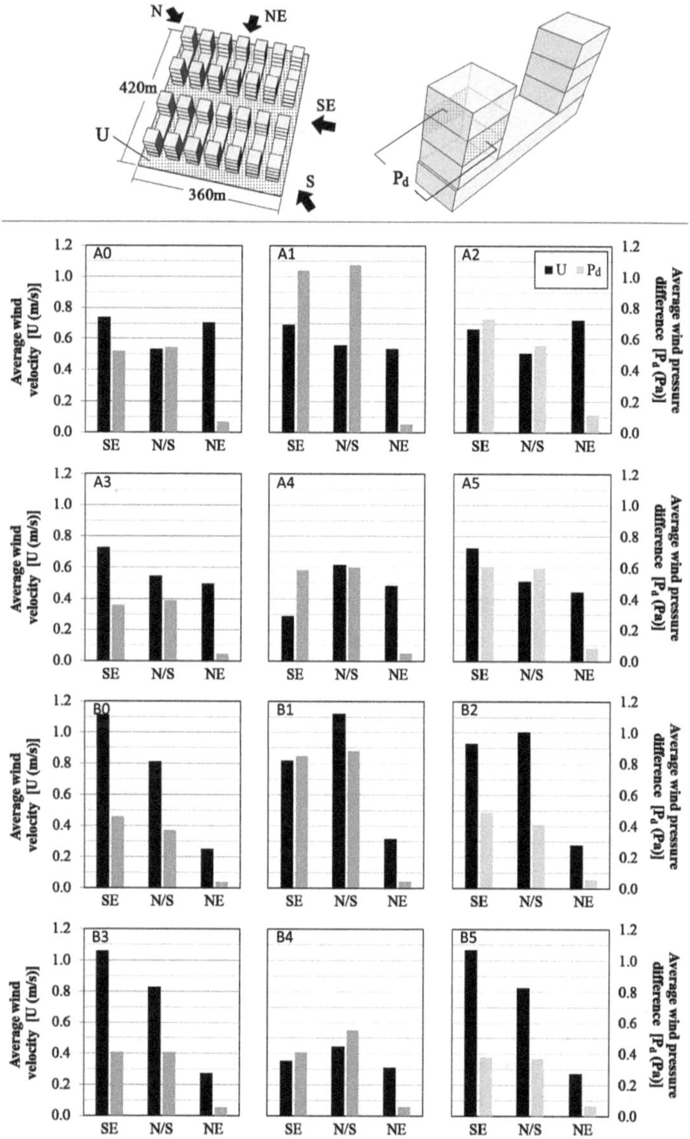

Fig. 9.2 Average wind velocity at pedestrian level (U) and average wind pressure difference (P_d) for five strategies of adjusting breezeway network patterns

breezeways is effective in improving both pedestrian and building ventilation potentials for most wind directions. However, a wind-impermeable podium structure should be avoided because it makes the airflow at pedestrian level less sensitive to the change of breezeway densities.

Unequal Breezeway Width

The second strategy varies the equal breezeway widths of 20 m to alternated breezeway widths of 10 m and 30 m.

At pedestrian level, compared with equal breezeway widths (A0 and B0), the values of U in the unequal cases (A2 and B2) increase when wind directions are from N, S and NE while they decrease in the case of SE wind direction for both plot types. Similarly, as with the previous strategy, varying breezeway widths is less effective for buildings with podium structures. Regarding the indoor ventilation potential at higher building levels, this strategy also tends to increase P_d. However, opposite to the situation of pedestrian ventilation, the podiums of plot type A benefit the building ventilation potentials.

The results suggest that varying breezeway widths should be an effective strategy to improve overall urban ventilation. In fact, the existing long and deep street canyons in CBD have a significant stagnation effect on the passing airflow. By introducing unequal breezeway widths, on one hand, more evident channelling effects are produced in wider breezeways. On the other hand, the narrower breezeways hardly worsen their original poor ventilation whilst enhancing larger pressure differences between opposite facades.

Unequal Breezeway Spacing

The third strategy varies the original equal spacing between the central axis of breezeways of 50 m to unequal alternated spacing of 42 m and 62 m. As a result, the plot and building widths are varied whilst breezeway widths remain constant.

At pedestrian level, with unequal breezeway spacing (A3 and B3), wind conditions are slightly worse than those in baseline models in most cases especially when wind is parallel to breezeways (NE and SE). For the upper levels, the situation is even worse. A more dramatic reduction of P_d is obtained for buildings with podium structures in all wind directions.

Overall, varying breezeway spacing is not recommended. Natural ventilation at both pedestrian and upper levels become worse for both plot types with most wind directions. The poor performance of the strategy is probably due to the introduction of wider building facades, which generate more stagnant areas near their leeward sides.

Staggering Blocks

The fourth strategy staggers the building blocks inside equal sized plots. In this situation, the staggering configuration has SE orientation.

For pedestrian ventilation, the staggered patterns (A4 and B4) perform much worse than the baseline cases that have aligned patterns. For most wind directions, U decreases to less than half of its original values. However, in the case of building ventilation potentials, staggered patterns create larger pressure differences especially for the case without podium (plot type B). This trend is more evident when wind comes from an oblique direction (N and S).

Overall, the strategy of staggering building blocks weakens the ventilation at pedestrian level whilst it increases the ventilation potentials of upper levels. Therefore, considering that its negative effect on pedestrian ventilation is much larger than its positive effect on building ventilation, this strategy should not be recommended unless further refinement of the staggering arrangement, especially by minimising wind obstruction at ground level, is investigated.

T-Shaped Breezeway Junctions

The fifth strategy replaces two X-shaped junctions by T-shaped junctions whilst breezeway width remains the same as in the baseline cases.

At pedestrian level, the creation of alternate T-shaped junctions (A5 and B5) hardly affects the wind performance compared with the baseline cases. The only obvious difference is that a significant reduction of U is obtained for plot type A when wind direction is NE. At upper levels, however, the T-shaped junctions increase P_d in every wind direction for plot type A whilst slightly reducing P_d when wind is from SE, N and S for plot type B.

According to the results, when a small number of T-shaped junctions is introduced among X-shaped junctions, urban ventilation conditions tend to remain unchanged or even be slightly weakened when BCR is small (plot type B). In comparison, for patterns with higher BCR and larger building bulks, T-shaped junctions tend to increase the building ventilation potentials.

Comparison and Analysis of the Strategies

The Best Compromise Between Pedestrian and Building Ventilation

As mentioned in the earlier sections, wind conditions at both pedestrian and upper building levels should be considered simultaneously in order to have a comprehensive evaluation of each strategy. Table 9.1 compares the effect of the five strategies on natural ventilation and thermal comfort.

From the results of this study, strategy 2 is considered to be the best compromise between pedestrian and building ventilation potentials. Varying the widths of neighbouring breezeways significantly increases the values both of U (up to 136.4 %) at pedestrian level and P_d (up to 69.6 %) for indoor ventilation potentials in comparison with the baseline patterns. These figures are top amongst all strategies. As Ramponi et al. (2015) reported, the introduction of wider breezeways should be the main reason for better ventilation. Additionally, strategy 1 should also be considered as a

Table 9.1 Effect of five strategies in terms of natural ventilation and thermal comfort in comparison with baseline cases

Strategy	*Percentage change of pedestrian ventilation*		*Percentage change of building ventilation*		*Percentage difference of outdoor area with thermal comfort*	
	Type A (%)	*Type B (%)*	*Type A (%)*	*Type B (%)*	*Type A (%)*	*Type B (%)*
1: Reducing breezeway densities	−24.1 to 5	−26.7 to 37.9	−21 to 99.6	−16.6 to 23.6	−26.2 to −2.6	−19.2 to 27.1
2: Varying breezeway widths	−10.1 to 2.4	10.8 to 136.4	2.3 to 69.6	6.3 to −46.3	−20.7 to −6.9	−9.8 to 28.7
3: Varying breezeway spacing	−29.7 to 2.5	−5 to 8.9	−28.5 to −31.7	−10.8 to 37.1	−37.4 to −3.5	−1.1 to 16
4: Staggering building blocks	−61.0 to 15.6	−68.4 to 24	−25.7 to 12.4	−10.9 to 48.6	−42.7 to 9.7	−62.8 to 3.4
5: Inserting T-shaped junctions	−36.6 to −3.5	−4.3 to 8.7	9.6 to 22.3	−17.6 to 69.9	−39 to −9.5	−0.9 to 13.1

very effective strategy for improving urban ventilation. These generate up to 37.9 % and 99.6 % increases of U and P_d, respectively. The figure of P_d in this strategy is even higher than that in strategy 2. However, it should be noted that decreasing breezeway density also leads to higher floor areas and deeper floor plans in each building unless further design refinement is applied. Deep floor plans may worsen the cross-ventilation effects inside buildings due to the probability of obstructive partition walls. Therefore, in these cases, higher P_d may not necessarily lead to good indoor ventilation.

On the contrary, strategy 3 and 4 are the worst. For instance, strategy 3 generates up to 29.7 % and 31.7 % reductions of pedestrian and building ventilation potentials, respectively. In fact the negative effects of varying the breezeway spacing are more serious than what the wind pressure data indicates. This is again due to the increase of building floor areas in larger plots. Strategy 4 leads to the most significant reductions of all. Over 60 % decrease of U is seen for both plot types when aligned building blocks are replaced by staggered building blocks. Though the strategy improves the building ventilation potentials with certain wind directions, it cannot compensate the negative effects on pedestrian ventilation. The result is consistent with Lin et al. (2014), who had reported that the staggered pattern gives lower pedestrian ventilation efficiency than the aligned pattern under the same conditions. They attributed this to the greater drag force produced by the staggered pattern, which might also explain these results.

Finally, strategy 5 should not be rejected but careful considerations should be taken due to the contradictory effects between pedestrian and indoor ventilation potentials. The use of T-shaped junctions decreases pedestrian ventilation whilst it significantly improves the building ventilation potentials. This means that other design variants with the insertion of T-shaped junctions should be further assessed to achieve an improvement of pedestrian ventilation whilst keeping the positive performance in terms of pressure differences and indoor ventilation potential. However, based on the findings, deep floor plans should be avoided for more efficient cross ventilation inside the buildings.

The Influence of Plot Types

An important consideration for urban planners is that breezeway network patterns and building typologies should be analysed in an integrated way. Both the effects of the evaluated strategies and of plot types on urban ventilation are significant.

Comparing the effects of plot types A and B (Table 9.1), at pedestrian level, buildings without podium structures tend to have better pedestrian ventilation regardless of the strategy applied. The largest difference amongst plot types with respect to the baseline cases can be seen when strategy 2 is applied, where the value of U is up to 136.4 % higher for plot type B and only up to 2.5 % higher for plot type A. For the other strategies, the difference is still evident. For example, the negative effects observed from strategies 3 and 5 on pedestrian ventilation is less pronounced for plot type B without podium structures.

At upper levels, plot type B also has better building ventilation compared to plot type A when strategies 3, 4 and 5 are applied. The most extreme difference between plot types is found when strategy 3 is applied, where type A experiences large reductions of P_d whilst increments of up to 37.1 % are obtained for type B. The serious reduction of P_d for buildings on podiums is mainly due to collisions of airflow caused by the unequal alternated building facades. However, when strategies 1 and 2 are applied, plot type A performs better than type B. The poor performance of plot type B is the result of the introduction of wider breezeways which disturb the original strong channelling effects along southeast-northwest axis. On the other hand, the wider breezeways in type A enhance the air ventilation along northeast-southwest axis.

To strike a balance between pedestrian and building ventilation potentials, buildings with large podium structures should be avoided. Though the plot with podium has good building ventilation at above-podium levels in some of the cases, it significantly worsens the pedestrian wind environments unless the podium becomes more permeable to the wind flow. This is consistent with the recommendations given by Ng (2009; 2010).

The Influences of Five Strategies on Outdoor Thermal Comfort

One of the major contributions of urban ventilation in Singapore's high-density areas is the improvement of thermal comfort conditions. By using the five strategies related to breezeway network patterns, different degrees of thermal comfort can be obtained. In this study, the proportion of urban areas at pedestrian level that obtain the minimum wind speed of 0.8 m/s is used as the indicator to explain the effect of the five strategies on outdoor thermal comfort. For indoor thermal comfort, further definition of interior spaces is needed to predict indoor air velocities.

Overall, due to the weak wind conditions in Singapore, thermal comfort for outdoor areas cannot be fully achieved by adjusting breezeway network patterns whilst keeping the high GPR. However, by applying a proper breezeway network pattern, building orientation and plot type, outdoor areas achieving thermal comfort can considerably increase. According to the data in Table 9.1, strategies1 and 2 obtain the largest increase of over 27 %; hence achieving over 50 % and 70 % outdoor areas with thermal comfort in total, respectively. These results confirm they are the most effective strategies in this study as discussed in section "The Best Compromise Between Pedestrian and Building Ventilation". Although strategy 3 can also obtain a high increase of comfort areas (up to 16 %), it cannot be considered as a good strategy for the following reasons: first, it gives lower average wind velocities than the baseline cases, meaning the remaining areas have very poor thermal conditions; and second, its low values of P_d and wide building blocks with deep floor plans should result in relative poor indoor thermal conditions.

BCR becomes a good indicator when the breezeway network pattern is kept constant. When BCR decreases from 40 % (plot type A) to 20 % (plot type B), urban areas that achieve thermal comfort can increase by 10–40 % in most breezeway network patterns. The only exception is seen when strategy 4 is used. This phenomenon further suggests that although staggered building blocks may be used for higher ventilation potential at building levels, the design of the towers should be decoupled from the ground level by providing more porous and linear breezeways at the pedestrian level.

Conclusions

This chapter explores the effects of a number of strategies related to breezeway network patterns on optimising urban ventilation and urban thermal comfort in a high-density and high-rise district in Singapore. The results suggest that the best improvement on urban ventilation at pedestrian and building levels is achieved by varying the breezeway widths in an alternated way. The five strategies were tested on two plot types, considering plot type B (without a podium structure) as the preferable one for most of the applied strategies. Further investigation is needed to evaluate the impact of a number of modifications on the models representing the best strategies as well as on plot types. A more promising compromise between pedestrian and indoor ventilation is expected to be achieved by adjusting breezeways dimensions and building forms.

The study of breezeway network patterns is an indispensable complement to the existing studies on urban ventilation. First, it offers a method of obtaining better urban ventilation without sacrificing density indicators in high-density areas in the tropics. Second, it gives guidance for planners to distribute plots with different land use. In fact, when road and landscape planning is conducted, planners should consider not only the traffic situations and landscape aesthetics, but also the airflow patterns; especially in hot and humid climates. A poorly ventilated breezeway network pattern may have a long-term negative effect since it is difficult to adjust in subsequent urban developments. To turn the dream of carbon neutral cities into a reality, whilst providing thermal comfort to inhabitants, urban ventilation should be placed at the forefront of urban policies and planning, especially in the tropical Asian cities.

References

Blocken, B., Stathopoulos, T., & van Beeck, J. P. A. J. (2016). Pedestrian-level wind conditions around buildings: Review of wind-tunnel and CFD techniques and their accuracy for wind comfort assessment. *Building and Environment, 100*, 50–81.

Buccolieri, R., Sandberg, M., & Di Sabatino, S. (2010). City breathability and its link to pollutant concentration distribution within urban-like geometries. *Atmospheric Environment, 44*(15), 1894–1903.

Building and Construction Authority. (2010). *Green building planning and massing*. Singapore: Building and Construction Authority.

Building and Construction Authority. (2012). *Certification standard for new buildings, BCA green mark (Version 4.1)*. Singapore: Building and Construction Authority. https://www.bca.gov.sg/GreenMark/others/GM_Certification_Std2012.pdf.

Chatelet, A., Fernandez, P., & Lavigne, P. (1998). *Architecture Climatique. Une contribution au développement durable 2*. Aix-en-Provence: Edisud.

Cheshmehzangi, A., Zhu, Y., Li, B. (2010). Integrated urban design approach: sustainability for urban design. *Proceedings for ICRM 2010, 5th International Conference for Responsive Manufacturing in China*. Ningbo.

Franke, J., Hellsten, A., Schlünzen, H., & Carissimo, B. (2007). *Best practice guideline for the CFD simulation of flows in the urban environment, COST 732. Quality assurance and improvement of microscale meteorological models*. Brussels: COST Office.

Georgakis, C., & Santamouris, M. (2005). Wind and temperature in the urban environment. In C. Ghiaus & F. Allard (Eds.), *Natural ventilation in the urban environment: Assessment and design* (pp. 81–102). London/Sterling: Earthscan.

Givoni, B., Khedari, J., Wong, N. H., Feriadi, H., & Noguchi, M. (2006). Thermal sensation responses in hot, humid climates: Effects of humidity. *Building Research and Information, 34*(5), 496–506.

Hong Kong Planning Department. (2005). *Feasibility study for establishment of air ventilation assessment system*. Final report, Hong Kong.

Kubota, T., Miura, M., Tominaga, Y., & Mochida, A. (2008). Wind tunnel tests on the relationship between building density and pedestrian-level wind velocity: Development of guidelines for realizing acceptable wind environment in residential neighborhoods. *Building and Environment, 43*(10), 1699–1708.

Launder, B. E., & Spalding, D. B. (1974). The numerical computation of turbulent flows. *Computer Methods in Applied Mechanics and Engineering, 3*(2), 269–289.

Li, B., Liu, J., & Li, M. (2013). Wind tunnel study on the morphological parameterization of building non-uniformity. *Journal of Wind Engineering and Industrial Aerodynamics, 121*, 60–69.

Lin, M., Hang, J., Li, Y., Luo, Z., & Sandberg, M. (2014). Quantitative ventilation assessments of idealized urban canopy layers with various urban layouts and the same building packing density. *Building and Environment, 79*, 152–167.

National Environment Agency. *Record of climate station mean, weather statistics 2016*. http://www.nea.gov.sg/weather-climate/climate/weather-statistics

Ng, E. (2010). *Designing high-density cities: For social and environmental sustainability*. London/Sterling: Earthscan.

Ng, E. (2009). Policies and technical guidelines for urban planning of high-density cities – Air ventilation assessment (AVA) of Hong Kong. *Building and Environment, 44*(7), 1478–1488.

Oke, T. R. (2002). *Boundary layer climates*. Abingdon: Routledge.

Rajagopalan, P., Lim, K. C., & Jamei, E. (2014). Urban heat island and wind flow characteristics of a tropical city. *Solar Energy, 107*, 159–170.

Ramponi, R., Blocken, B., Laura, B., & Janssen, W. D. (2015). CFD simulation of outdoor ventilation of generic urban configurations with different urban densities and equal and unequal street widths. *Building and Environment, 92*, 152–166.

Stocker, T. F., Qin, D., Plattner, G.-K., Tignor, M., Allen, S. K., Boschung, J., Nauels, A., et al. (2014). *Climate change 2013: The physical science basis*. Cambridge/New York: Cambridge University Press.

Tominaga, Y., Mochida, A., Yoshie, R., Kataoka, H., Nozu, T., Yoshikawa, M., & Shirasawa, T. (2008). AIJ guidelines for practical applications of CFD to pedestrian wind environment around buildings. *Journal of Wind Engineering and Industrial Aerodynamics, 96*(10–11), 1749–1761.

Tsutsumi, J., Katayama, T., & Nishida, M. (1992). Wind tunnel tests of wind pressure on regularly aligned buildings. *Journal of Wind Engineering and Industrial Aerodynamics, 43*(1), 1799–1810.

United Nations, Department of Economic and Social Affairs, Population Division. (2017). *World Population Prospects: The 2017 Revision, Key Findings and Advance Tables*. New York: United Nations. https://esa.un.org/unpd/wpp/Publications/Files/WPP2017_KeyFindings.pdf

Wong, N. H., & Chen, Y. (2005). Study of green areas and urban heat island in a tropical city. *Habitat International, 29*(3), 547–558.

Wong, N. H., & Chen, Y. (2008). *Tropical urban heat islands: Climate, buildings and greenery*. Abingdon: Taylor & Francis.

Wong, M. S., Nichol, J. E., Pui Hang To, & Wang, J. (2010). A simple method for designation of urban ventilation corridors and its application to urban heat island analysis. *Building and Environment, 45*(8), 1880–1889.

Yang, W., Wong, N. H., & Jusuf, S. K. (2013). Thermal comfort in outdoor urban spaces in Singapore. *Building and Environment, 59*, 426–435.

Zaki, S. A., Hagishima, A., & Tanimoto, J. (2012). Experimental study of wind-induced ventilation in urban building of cube arrays with various layouts. *Journal of Wind Engineering and Industrial Aerodynamics, 103*, 31–40.

CHAPTER 10

'Summoning' Wind for Urban Cooling: Urban Wind Corridor Projects in China

Dong-Li Hong and Shiuh-Shen Chien

Introduction: Governing the Urban Climate

Over the past century, urbanisation and industrialisation have had a dramatic global impact on climate. Urbanisation has drastically altered landscapes, covering huge areas in concrete and asphalt surfaces. Human activity has released tremendous amounts of pollutants and waste heat into urban atmospheres, resulting in air pollution, smog and the urban heat island (UHI) effect. The management of climate change has emerged as a major challenge for architects, planners, urban scholars, city officials, private industry and civil society (Emmanuel 2005).

Urban-planning studies and practices seek to respond to these challenges. Under the state-centred developmentalism approach to economic growth, urban planning is very focused on land use changes and industrial zone planning to promote investment and production (Chien 2008; Wei and Leung 2005). Urban planning later began to emphasise environmentalism to remedy industrial land and water pollution created by unregulated economic activities. Recent years have seen progress on urban responses to climate change such as heavy flooding, air pollution, smog and UHI.

D.-L. Hong • S.-S. Chien (✉)
National Taiwan University, Taipei, Taiwan

A. Cheshmehzangi, C. Butters (eds.), *Designing Cooler Cities*, Palgrave Series in Asia and Pacific Studies,
https://doi.org/10.1007/978-981-10-6638-2_10

In terms of planning strategies to manage urban climate and weather environments, most existing literature focuses on adaptive solutions, including greenification projects to create more urban parks and green roofs, and sponge city projects that increase capacity to deal with heavy rain and flooding. Actively interventional solutions are also explored, including a rain-enhancement experiment conducted in Tai Lake in China's Jiangsu Province for the reduction of local temperatures; thus also inhibiting cyanobacteria growth which can cause eutrophication and damage water quality. Results from the Tai Lake experiment have shown that air temperature decreased by 1–2 °C with the rain enhancement by weather modification (Wang and Chen 2015). However, while the experiment demonstrated the potential to control short-term weather, it had no long-term effects on the urban climate problems such as managing directions of wind for urban cooling or controlling smog.

Wind is a key factor in determining urban thermal environments (Morris et al. 2001). It boosts the circulation of air and heat and decreases apparent temperatures for human comfort. Projects to establish urban ventilation corridors to improve wind flow through cities are under development in Asian countries including Japan (Ashie et al. 2006; Akashi 2008; Chen et al. 2009) and South Korea (Cha et al. 2007). China is developing large-scale ventilation corridor projects, with feasibility studies completed in at least a dozen cities (see Table 10.1), with land-use and building-regulation policy initiatives currently in process in Guiyang, the provincial capital of Guizhou, located in south of China. These provide an opportunity to scrutinise how planning strategies can be implemented for 'summoning wind'.

The rest of this chapter is divided into three main sections. The first section briefly describes the role of wind planning in architecture and urban-planning-studies literature and the use of various computer models in wind planning. The second section discusses the promotion of urban ventilation corridors in China, with Guiyang's urban ventilation corridor plan as a central case study. The third section discusses social tensions and other issues arising from such projects.

Urban Wind Planning and Computational Modelling

Wind has long been recognised as a critical element of planning the urban built environment. The ancient philosophy of '*feng shui*', literally translated as 'wind-water', can be traced back to the Zhou dynasty of China (about 1050–250 BC) and reflects traditional Chinese insights into spatial

Table 10.1 Selected urban wind corridor studies and planning projects in China

City	*Project*	*Research period*
Direct related to wind ventilation corridors		
Wuhan	Study of planning and management of Wuhan urban air path [Wuhan *chengshi fengdao guihau guanli yanjiu*]	2012–2013
Changsha	Technical guide for urban air ventilation in Changsha city [Changsha *shi chengshi tongfeng guihua jishu zhinan*]	2010
	Making natural ventilation corridors in cities with high seasonal temperature variation [*Xiare dongleng diqu chengshi ziran tonggeng kangdao yingzao moshi yangjiu*]	2010
Hangzhou	Urban air path study [*Chengshi tongfeng langdao guihua yanjiu*]	2013
Guiyang	Study of urban ventilation corridor planning methods based on Guiyang [*Chengshi tongfengdao guihus sheji fangga yanjiu*]	2013
	Layout plan of ventilation corridor below about 200 meters in downtown Guiyang [Guiyang *shizhongxin chengqu* 200 *gongchi yixia jindi tongfeng langdao buju guiahua*]	2013
Overall wind environment planning studies		
Wuhan	A study of the relationship between the layout of urban building planning and its surrounding climate environment [*chengshi jianzhu guihua buju yu qihoiu guanxi yangjiu*]	2005
Langfang	Langfang's wind environment [Langgan *chengshi fenhuanjing de chengshi guihua yangjiu*]	2011
Xi'an	Planning for ecological isolation mechanisms in the Xi'an region [*xian shiyu shengtai geli tixi guihua*]	2013
Shenyang	Green space control in urban spatial structure planning in Shenyang [Shenyang *chengshi jiegouxing ludi kongzhi guihua*]	2014
Nanjing	Air pollution protection plan in Nanjing [Nanjing *shi daqi wuran fagzhi xingdong jihua*]	2015

Source: Adopted from Ren et al. (2014)

arrangements designed to strike a harmonious and healthy balance between built and natural environment. As urban climatology emerged in the early twentieth century, planners started noticing the effects of urban land-use and landscape patterns on the micro-climatology (Hebbert 2014). However, while some factors such as the spatial pattern of the UHI effect were easily captured and measured, other variables, like wind, were hard to observe and analyse thoroughly due to its rapidly changing and non-linear characteristics (Mills 2014).

Without proper research and design tools, urban planners can only consider and use climate information unsystematically (Eliasson 2000) either to depict existing conditions or give general recommendations for microclimate analysis and planning (Erell 2008; Ebrahimabadi et al. 2015). Thus, urban climatology knowledge, especially regarding wind, had very limited impact on planning processes in the twentieth century. In 1953, a pioneering urban trial on wind was done in Stuttgart, a German industrial city with problems of air pollution due to a combination of its valley location and low wind speeds. Guided by considerations of the natural wind paths, the urban plan in Stuttgart connected green spaces and discouraged development that would disrupt local air circulation (Hebbert and Webb 2012). The Stuttgart case demonstrated the possibility of combining the concern of wind and urban planning instructions, though hard without proper tools and knowledge to bridge the gap between urban climatology and planning processes. It was not until the development of modern computational modelling of micro and meso scale wind, which allows for more precise forecasts, that the possibility of wind planning emerged. Two kinds of modelling can be identified.

The first is small-scale modelling that works to predict air circulation surrounding one building or a group of buildings. A related issue is skyscraper-affected airflows. Improvements to construction techniques allow skyscrapers to be narrower and taller, and these structures can create street level wind hazards through downdraughts and wind tunnels (Liu 1990; Wells 2005). The former happens where moving air hits a building and is forced up or down, creating strong winds at street level. The latter refers to wind acceleration created by air passing through narrow channels between skyscrapers and nearby buildings. The complexity of these phenomena requires micro-climate analysis during architectural programming and construction. Researchers have used small-scale models to capture the flow and diffusion patterns of pollutants and energy in urban environments (Mfula et al. 2005). The computational fluid dynamic (CFD) model, for example, is commonly used to simulate and demonstrate small-scale wind environments. CFD applies fluid mechanics along with numerical analysis and algorithms with a set of physical models to simulate better the interaction and diffusion of liquids, gases and heat within boundary conditions. Due to its site-specific characteristics and limited computational resources, CFD and other small-scale models are often used to enhance the accuracy of monitoring and simulations at street and district levels (Cheshmehzangi 2016). For example, Akashi (2008) discussed changes in wind environment and mitigation of UHI in Tokyo after deconstructing a wall-shaped harbour-front building. Similarly, Ashie et al. (2006) discussed the UHI within a 10 km^2 area in central Tokyo using CFD. Although the small-scale model can simulate wind environments for architectural design, its computationally intensive nature makes it difficult to apply to city-level planning.

The second category of model focuses on simulating large-scale air flows and how they respond to land-use patterns, based on GIS data. These models divide urban surfaces into grids and assign each block a value for 'roughness' to calculate large-scale air flows in urban environment (Grimmond and Oke 1999; Mirzaei and Haghighat 2010; Ng et al. 2011; Gal and Unger 2009). Using simplified assumptions and numerical approximations to estimate roughness parameters (including zero-plane displacement height, roughness length, plan area density, frontal area index, average height weighted with frontal area, etc.) the large-scale model can provide appropriate simulations for urban scale wind flows. Previous efforts have combined these models with meteorological models to improve the spatial resolution of wind flow simulations, providing additional details for planning urban wind corridors (Chen et al. 2011; Hsieh and Huang 2016; Zhou et al. 2015).

From around 2005, cities in China have applied different models to the development of urban wind corridors (Ren et al. 2014). These efforts raise questions including: How can computational modelling results be effectively used for wind planning?; What are appropriate planning strategies to regulate land use patterns and building designs for creating urban ventilation corridors?; What social tensions take place beyond the wind corridor plans that attempt to improve the urban thermal environments? In the following section, we examine these issues in the context of a wind corridor being planned in the City of Guiyang.

Case Study Analysis: Wind Corridor Plan in Guiyang City

Planning of urban wind corridors first emerged as an important issue in China in the early-2000s, with institutions and administrations in various cities researching the feasibility of achieving aims including mitigation of air pollution, smog, and UHI (see Table 10.1). At least a dozen cities, including, Beijing, Shanghai, Jinan, Fuzhou, Huangshi, Wuhan, Hangzhou, Guangzhou, Zhengzhou, Xian and Hefei are either doing wind ventilation corridor feasibility studies or proposing wind ventilation corridor projects.[1]

Guiyang helped pioneer wind corridors through establishment of a relatively substantial and comprehensive development plan. With a population of around 4.5 million, Guiyang is located at the eastern edge of the

Yungui Plateau and consists largely of mountainous and hilly land with an average elevation of around 1100 m above sea level. The high altitude and low latitude results in relatively calm weather patterns with an average annual temperature of 15 °C, with southern winds in summer, shifting to north-east in winter. However, this good climatic environment is now facing certain challenges due to Guiyang's relatively compact urban built environment, which significantly reduces local wind speeds to an average of 2.49 m/s in some areas, such as the old city centre. It is, therefore, unable to 'naturally' tackle smog pollution and urban heat environmental problems. The city government started a project in cooperation with Huazhong University of Science and Technology for a wind corridor plan in order to ensure more comfortable, healthy conditions for the citizens.

Guiyang's planning used a meso-scale meteorological model (the Advanced Research Weather Research and Forecasting Model, WRF-ARW) to simulate the wind environment, incorporating additional data such as the average height of city blocks and street widths to adjust the single-layer urban canopy model to capture better the detailed local characteristics (Zhou et al. 2015). The model not only allows researchers to simulate the current wind environment, but also make adjustments to examine the impact of various wind corridor parameters. Simulation results were incorporated into the 'Layout Plan of Ventilation Corridor Near 200 Meters in Downtown Guiyang' document, submitted by the Guiyang Urban and Rural Planning Committee in 2014, including: (a) establishment of three level wind corridors; and (b) regulations and limitations of corridor zones (Fig. 10.1). This plan was then incorporated into the 'Guiyang Blue Sky Protection Project (2014–2017) (Guiyang *lantian baohu jihua*)' announced by the Guiyang city government as a key strategy to mitigate urban air pollution and the urban heat island (UHI) effect.

Establishment of Three-Level Wind Corridors

In addition to the planning foundation provided by the computer-simulation results, the plan also considered other local spatial factors and conditions. Natural terrain features, such as, valleys were selected for use in the Level-1. Existing transportation infrastructure, including railways, highways and main roads, with orientations matching prevailing wind, are used as ventilation corridors. Third, urban spaces, such as parks, were designed to improve air circulation. Fourth, regulations were implemented to promote non-uniform building heights to create air paths for ventilation corridors.

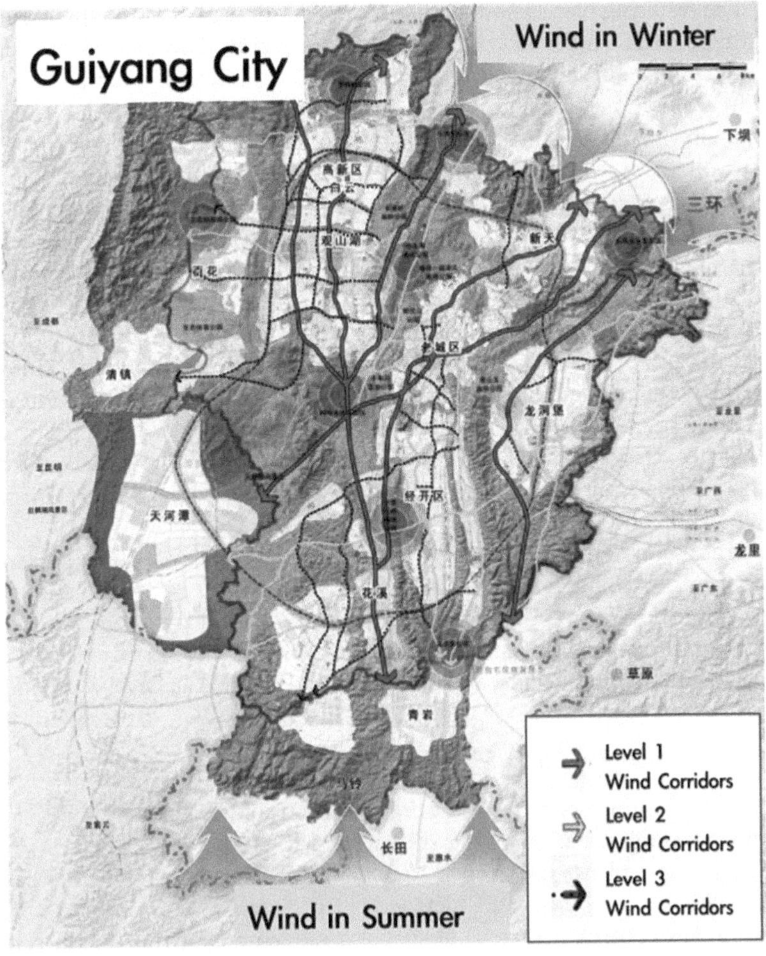

Fig. 10.1 Planning of three wind corridors in Guiyang (Source: Adopted from Guiyang Urban and Rural Planning a Design Institute 2014)

Finally, old town regeneration and relocation of industrial zones are regarded as another two opportunities to change large-scale land use patterns in order to install ventilation corridors in the urban transformation process.

The resulting plan included multiple corridors at each of the three levels: 3 for Level-1, 16 for Level-2, and 30 for Level-3. The Level-1 corridors were established parallel to the main direction of Guiyang's prevailing winds with a corridor width ranging from 100 to 300 metres. The Level-1 corridors linked existing green belts and water bodies including rivers, reservoirs, and wetlands to form a wide and linear natural wind path.

Connected to the Level-1 corridors are 16 Level-2 corridors ranging from 50 to 300 metres width to form a ventilation network for the city. In addition to natural wind paths, the Level-2 corridors incorporate various artificial paths, including railways, highways and major roads. These artificial infrastructures create continuous channels for air flow through the city, and their intersections with the Level-1 corridors form an effective air diffusion network. Human activity within these channels is one of the main sources of air pollutants and waste energy. A network of well-connected wind corridors can decrease the density of air pollutants emitted by cars, and lower the thermal effect from concrete or asphalt pavements. The combination of Levels-1 and-2 forms the main wind corridor system for diffusing pollutants and heat.

In addition to the Level-1 and Level-2 corridors, 30 Level-3 corridors with narrower controlling areas of 30–100 metres are planned in the zones surrounding the major corridor network for further wind diffusion. Areas that are roughly parallel to the main orientation of the main wind corridor system with lower building heights and densities were assigned as Level-3 corridors.

Urban Planning Control for the Wind Corridor

Feasibility studies for the Guiyang project require its integration into China's overall urban planning system, which is composed of three kinds of statutory plans: masterplan [*zongtixing guihua*], regulatory plan [*kongzhixing guihua*], and detailed plan [*xiangxixing guihua*]. The masterplan controls the general spatial arrangement of the overall urban area, and serves as the foundation for urban land-use and development. Based on the masterplan, the regulatory plan provides relevant regulations to constrain various land use patterns, including the intensity of development, the limit of building coverage ratio and floor area ratio,[2] and permissible building heights. Finally, a detailed plan is established for each building project to guide design and construction.

The paths of the corridors for all three levels are first identified as part of the masterplan. Urban planning strategies and development constraints are then formulated for the regulatory and detailed plans to regulate and

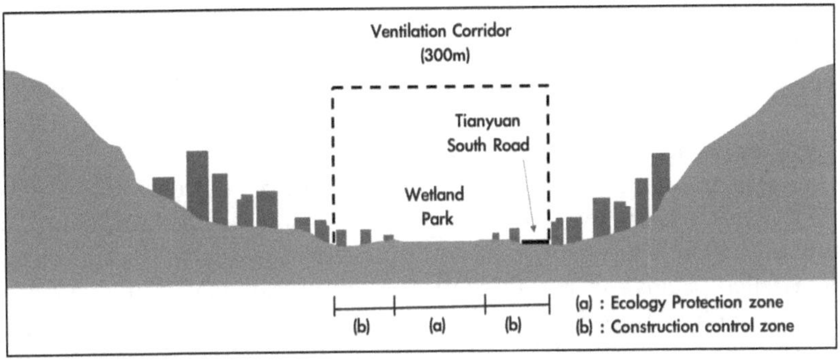

Fig. 10.2 Section view of the Southern Hua Stream Level-1 ventilation corridor (Source: Author compiled data, and adopted from Guiyang Urban and Rural Planning & Design Institute 2014)

control areas included in the wind corridor control zones. The range of the zones follows the widths of the corridor levels. In addition to the width of the control zones, constraints are made on development intensity, particularly for building heights. For example, within the Southern Hua Stream Level-1 ventilation corridor, to be established along with a 300-metre-wide wetland park, new buildings are restricted to a height of 12 m (Fig. 10.2). By contrast, buildings in the construction control zone of the Xintian Boulevard Level-2 ventilation corridor can be as high as 36 m.

In addition to regulatory plan restrictions, effective ventilation requires regulation of building forms and shapes (e.g. minimising large slab-style/longitudinal buildings). Simulations suggest that implementation of the proposed regulations can increase wind speeds in central Guiyang by 10 % (Zhou et al. 2015), with the most significant increase (up to 1.5 m/s) in the high-density old city area, which also features the greatest concentration of pollutants and the most significant heat island. Simulations also show the proposed corridors can lower urban temperatures by an average of 0.3 °C, and 0.65 °C in the old city area.

Challenges of Urban Wind Planning Facing China

Cities in China have devoted considerable resources to studying the feasibility of urban wind planning in general and wind corridors in particular. Guiyang is pioneering the development of relevant land-use regulations and planning laws. The experience of Guiyang and other Chinese cities can provide a useful reference for the impact of urban wind planning on local populations and social conditions, and the challenges facing urban authorities in instituting required regulations. In this respect, four kinds of challenges are particularly identified.

First, at the policy-implementation level, China's current planning system features a gap between regulatory planning and detailed planning. The regulatory plan is too generic to guide construction at the block level, but the detailed plan is too specific to provide overall adequate building design patterns for wind corridors. An intermediate step for urban design that requires examining principle spatial patterns and arrangements before the announcement of a detailed plan must be introduced to fill this gap. However, urban planning policy makers in China are still debating how to develop the urban planning regulation structure and it may take a long time before urban design officially incorporates urban wind corridor plans.

Second, wind planning and climate change research rely heavily on computational modelling. Although climate-related modelling has made substantial progress in recent years, projections still suffer from a high degree of uncertainty. At least three inter-related dimensions complicate improving projection accuracy, including: (1) In theory, the urban roughness index should include detailed information about most buildings and objects along the frontal area. But in the context of rapidly changing urban environments, as in China, keeping an updated urban roughness index can be time and energy consuming; (2) Improved forecasting of surface wind directions, strengths and sources requires additional local basic and long-term meteorological data. Most cities lack sufficient weather monitoring resources; and (3) Additional indexes and data increase the computational loading required for modelling, which entails significant costs.[3]

Third, wind planning focuses on improving average seasonal or annual climatic conditions. However, public perception focuses on time and location-specific conditions, such as heavy smog. It may be reasonable for residents to question whether wind corridors are practicable in cities where the winds generally do not blow from a single general direction. Unfortunately, some urban leaders promote the concept of wind ventilation corridors because it is a trend rather than because the physical environment of their cities are suitable.

Fourth, effective implementation of any urban ventilation corridor plans requires land-use restrictions that either prohibit high-density development in specific areas or involve demolition of existing structures if necessary.

Certain possible results are particularly noted. Any restriction of high-density land use development is certainly against the financial profitability of land, which is a very basic motivation for local cadres' behaviour in the current political context (Lin 2011). On top of that, any demolition can often result in significant social disruption and resistance, which local leaders are keen to avoid. In addition, given the high degree of uncertainty that characterises current climate and wind pattern modelling, building wind ventilation corridors is seen as a passive solution. More sustainable solutions could focus on slowing economic development and reducing emissions. However, more sustainable solutions that cannot generate better gross domestic product are not attractive as priority policies to most local leaders in China.

Conclusion

Through the case of urban planning of the wind corridors in Guiyang city in China, this chapter illustrates a newly rising concern about wind in the practice of urban planning. With different scales of computer modelling, urban planners can both capture the patterns of air flow at both micro and meso scale, and also simulate the effects on wind speed and thus effects on temperature as well as air pollution, of different land-use patterns. Hence, computer modelling is implemented as a tool to bridge the urban climatology and the urban planning practices. However, the case of China also shows that the socio-political consequences of urban wind corridors needs to be further examined.

In a broader sense, the urban plan of wind corridors matches with the rising trend in many Chinese cities in terms of adaptation to climate change by introducing new concepts and technology. For instance, the concept of sponge city, referring to absorbing precipitation through permeable pavements, rain gardens and wetlands, has been promoted widely in many Chinese cities. The cases of sponge city and wind corridor certainly show that some Chinese local officials are willing to take urban climatology and lower density of land use into consideration in order to deal with the challenges of climate change and problems of urban heat. However, the introduction of these ideas is just beginning in China, and we need more research on how these climate change adaptation strategies are actually implemented without too seriously compromising concerns of local land finance.

Notes

1. News source: Beijing and other cities plans to introduce wind ventilation corridors to curb PM 2.5 (Ch. Beijing *deng duoge chengshi niguihua 'chengshi fengdao' shusan* PM2.5). https://read01.com/7B04Jz.html, accessed 31 December 2016; and Effectiveness of multiple cities planning for ventilation corridors to curb smog needs further reviews (Ch. *duochengshi mouhua 'chengshi fengdao' zhimai xiaoguo daigu*) http://www.1000plan.org/qrjh/article/56856, accessed 30 November 2016.
2. Building coverage ratio and floor-area ratio, two indicators for the land use area of buildings and volume and height of buildings, are a zoning guidance to prevent tall buildings from obstructing too much light and air.
3. News source: Urban ventilation corridors: wind passing through the city [*chengshi fengdao chuanchengfeng nengfou chuisan zhongguo de wumai*], http://www.storm.mg/article/97745, accessed 31 December 2016.

References

Akashi, T. (2008, September). Creating the 'wind paths' in the city to mitigate urban heat island effects—A case study in Central District of Tokyo. In CIB-W101 (Spatial Planning and Infrastructure Development) annual meeting, Dublin, Ireland

Ashie, Y., Tokairin, T., Kono, T., & Takahashi, K. (2006, April). Numerical simulation of urban heat island in a ten-kilometer square area of central Tokyo. Annual report of the earth simulator center, pp. 45–48.

Cha, J. G., Jung, E. H., Ryu, J. W., & Kim, D. W. (2007). Constructing a green network and wind corridor to alleviate the urban heat-island. *Journal of the Korean Association of Geographic Information Studies, 10*(1), 102–112.

Chen, H., Ooka, R., Huang, H., & Tsuchiya, T. (2009). Study on mitigation measures for outdoor thermal environment on present urban blocks in Tokyo using coupled simulation. *Building and Environment, 44*, 2290–2299.

Chen, F., Kusaka, H., Bornstein, R., Ching, J., Grimmond, C. S. B., Grossman-Clarke, S., et al. (2011). The integrated WRF/urban modelling system: Development, evaluation, and applications to urban environmental problems. *International Journal of Climatology, 31*(2), 273–288.

Cheshmehzangi, A. (2016). Multi-spatial environmental performance evaluation towards integrated urban design: A procedural approach with computational simulations. *Journal of Cleaner Production, 139*, 1085–1093.

Chien, S.-S. (2008). Isomorphism of local development policy: A case study of the formation and transformation of national development zones in post-Mao Jiangsu, China. *Urban Studies, 45*(2), 273–294.

Ebrahimabadi, S., Nilsson, K. L., & Johansson, C. (2015). The problems of addressing microclimate factors in urban planning of the subarctic regions. *Environment and Planning B: Planning and Design, 42*(3), 415–430.

Eliasson, I. (2000). The use of climate knowledge in urban planning. *Landscape and Urban Planning, 48*(1), 31–44.

Emmanuel, M. R. (2005). *An urban approach to climate-sensitive design: Strategies for the tropics.* New York: Taylor & Francis.

Erell, E. (2008). The application of urban climate research in the design of cities. *Advances in Building Energy Research, 2*(1), 95–121.

Gal, T., & Unger, J. (2009). Detection of ventilation paths using high-resolution roughness parameter mapping in a large urban area. *Building and Environment, 44*(1), 198–206.

Grimmond, C. S. B., & Oke, T. R. (1999). Aerodynamic properties of urban areas derived from analysis of surface form. *Journal of Applied Meteorology, 38*(9), 1262–1292.

Guiyang Urban and Rural Planning & Design Institute. (2014). *Layout plan of ventilation corridor near 200 meters in Downtown Guiyang.* http://www.planning.cn/view.php?id=364

Hebbert, M. (2014). Climatology for city planning in historical perspective. *Urban Climate, 10*, 204–215.

Hebbert, M., & Webb, B. (2012). Towards a liveable urban climate: Lessons from Stuttgart. In C. Gossop & S. Nan (Eds.), *Liveable cities: Urbanising world* (pp. 132–150). The Hague: International Society of City and Regional Planners. http://archive.ffue.org/PDF/Isocarp-review7-2011.pdf.

Hsieh, C. M., & Huang, H. C. (2016). Mitigating urban heat islands: A method to identify potential wind corridor for cooling and ventilation. *Computers Environment and Urban Systems, 57*, 130–143. https://doi.org/10.1016/j.compenvurbsys.2016.02.005.

Lin, G. C. S. (2011). China's landed urbanization: Neoliberalizing politics, land commodification and municipal finance in the growth of metropolises. *Environment and Planning A, 46*(8), 1814–1835.

Liu, H. (1990). *Wind engineering: A handbook for structural engineering.* Upper Saddle River: Prentice-Hall, A Pearson Education Company.

Mfula, A. M., Kukadia, V., Griffiths, R. F., & Hall, D. J. (2005). Wind tunnel modelling of urban building exposure to outdoor pollution. *Atmospheric Environment, 39*(15), 2737–2745.

Mills, G. (2014). Urban climatology: History, status and prospects. *Urban Climate, 10*, 479–489.

Mirzaei, P. A., & Haghighat, F. (2010). Approaches to study urban heat island—Abilities and limitations. *Building and Environment, 45*(10), 2192–2201.

Morris, C. J. G., Simmonds, I., & Plummer, N. (2001). Quantification of the influences of wind and cloud on the nocturnal urban heat island of a large city. *Journal of Applied Meteorology, 40*(2), 169–182.

Ng, E., Yuan, C., Chen, L., Ren, C., & Fung, J. C. (2011). Improving the wind environment in high-density cities by understanding urban morphology and surface roughness: A study in Hong Kong. *Landscape and Urban Planning, 101*(1), 59–74.

Ren, C., Yuan, C., Ho, C., & Ng, Y. (2014) A study of air path and its application in urban planning [Ch. *chengshi fenglangdao ji qi guihua yingyong*]. *Urban Plan Forum* [Ch. *chengshi guihua xuekan*], *3*, 52–60.

Wang, J., & Chen, Y. (2015). Effect analysis of artificial rain enhancement on prevention and control of Taihu Lake Cyanophyta in 2010 summer. *Chinese Agricultural Science Bulletin, 31*(14), 232–237.

Wei, Y. D., & Leung, C. K. (2005). Development zones, foreign investment, and global city formation in Shanghai. *Growth and Change, 36*(1), 16–40.

Wells, M. (2005). *Skyscrapers: Structure and design*. London: Laurence King Publishing.

Zhou, X., Chen, H., & Guan, Y. (2015) Study of urban ventilation corridor planning method based on a case study of Guiyang, China [Ch. *chengshi tongfengdao guihua sheji fangfa yanjiu- yi* Guiyang *shi weili*]. *Journal of Human Settlements in West China* [Ch. *xibu renju huanjing xuekan*], *6*, 13–18.

CHAPTER 11

District Cooling: A Key Solution for Hot Climate Cities

Chris Butters, Adzuieen Nordin, and Danny Tam Hong Khai

Introduction

This chapter addresses district energy solutions at the urban scale, principally though not uniquely for cooling. Whilst district heating (DH) systems are widespread, district cooling (DC) is less well known, although DC systems have existed for some years in both hot-dry and in hot-humid climates. There are major efforts today to spread awareness of DC, not least in Asia. After a brief overview of the development worldwide and scope of district energy systems, we focus on Malaysia, which is a leading example. We also highlight a little discussed conflict between the level of individual buildings versus that of urban energy planning. Which level should be prioritised and under what conditions?

In the rapidly growing cities of developing countries, millions are moving to dense urban environments, often of mediocre quality and with few or any energy-efficiency measures. Temperatures are typically 2–4 degrees hotter

C. Butters (✉)
University of Warwick, Coventry, UK

A. Nordin
Politeknik Ungku Omar of Malaysia, Ipoh, Malaysia

D.T.H. Khai
Shinryo Corporation, Kuala Lumpur, Malaysia

A. Cheshmehzangi, C. Butters (eds.), *Designing Cooler Cities*,
Palgrave Series in Asia and Pacific Studies,
https://doi.org/10.1007/978-981-10-6638-2_11

in inner city areas, and rising. The fraction of the world's 30 largest cities that lie in the tropics is forecast to rise from 40 % in 2000 to 60 % by 2025 (UN 2014). Typically, heat from vehicles accounts for between 20 % and 30 % of the anthropogenic sources of heat in large cities, whilst the energy use in buildings accounts for 60–75 % (Stewart and Kennedy 2015). The common cooling solution is still small-scale air conditioning (AC) units in individual buildings. Each one exhausts heat to the outdoors, hence heating up the city even more. Thus, ironically, *cooling* is itself one of the major causes of heat in the city. In addition, small AC units are inefficient, not always healthy, and are expensive to run. Larger urban scale solutions offer much higher technical efficiency as well as advantages as to energy sources, costs and management. And not least, district energy solutions offer one of the only ways to counteract the urban heat island (UHI) effect.

Better technology can help, but the fundamental challenge is to remove the sources of heat from the city. This is what DC systems offer; the unique feature of *reducing* UHI, thus impacting positively on environment as well as on comfort and public health. District cooling is a key to future urban planning and energy policy in hot climates.

District Energy Systems

The term district energy systems is applied quite broadly to solutions at a scale ranging from that of a housing neighbourhood or industrial park to that of entire city districts. Many provide electricity as well as cooling in combined heat and power (CHP) systems. They can be developed and run by the public or private sector, and can provide winter heating, summer cooling or both; many cities in quite cold climates now have a large summer cooling need too.

Space cooling is the largest energy need in buildings with modern comfort levels and is also (along with traffic) the fastest growing source of emissions and UHI in hot climate cities. In contrast to *electricity* supply, DC systems supply *low temperature energy*, which normally constitutes a larger fraction of end-use needs than the electricity needed for lighting and appliances. A key advantage is thus that the energy is supplied in the end-use form required. Electricity is as we know a high-quality energy carrier (exergy) which unless generated by renewables implies high greenhouse gas emissions.

With DC, the energy production and resultant emissions of waste heat are moved outside the city, removing both pollution and heat from the urban environment. On the downstream side, various heat sinks—the air, rivers, or the sea—can be used to dissipate the waste heat. Ecosystems will be significantly affected only where very large energy plants eject heat to a relatively small recipient. The challenges of DC are not primarily technical but ones of energy policy, urban planning and financing. DC requires standard technologies of pumps, piping and heat exchangers. Whilst many of today's DC systems operate with gas (Fig. 11.1), the energy source can also be switched in future from fossil fuels to renewables. In large systems, ice storage is often added, to reduce peak load demands and to exploit cheaper night time energy tariffs. Yet another advantage is that DC provides cooling without vibrations or noise as compared to many mechanical ventilation and cooling systems.

The first DH system dates back as far as 1877 in New York. The first DC was built in Colorado in 1930; in Europe, the first were built in Paris and Hamburg in the 1960s. An initial wave of interest in seawater cooling systems (SWAC) occurred in the 1970s (Argonne National Laboratory 1977). The focus then was energy security, before the concern for climate emissions. Interest waned as oil prices eased, although interest continued in locations such as California and Hawaii (Radspieler et al. 1999; State of Hawaii 2002). DH has been quite widespread in northern Europe for some 30 years. But UHI is becoming a major issue in colder climates too; a recent study of Paris found that city temperatures could increase by a further 2 °C if the amount of AC is doubled, as is projected (Munck et al. 2013). DC systems are few and far between. Sweden is an exception with around 25 % of cooling already supplied by DC (Dalin 2012), but in Europe as a whole, the market share of DC is under 2 % (European Commission 2015).

Natural Sources for District Cooling

Most conventional small-scale AC takes air at ambient outdoor temperature and cools it; but cooling becomes simpler if a natural source is available that has lower than ambient air temperatures. Even a few degrees lower temperature improves efficiencies significantly. Whilst many DC plants use conventional energy sources such as gas, including most of those in Malaysia, others exploit natural sources of cooler air or water where available. These include rivers, deep water bodies and the underground.

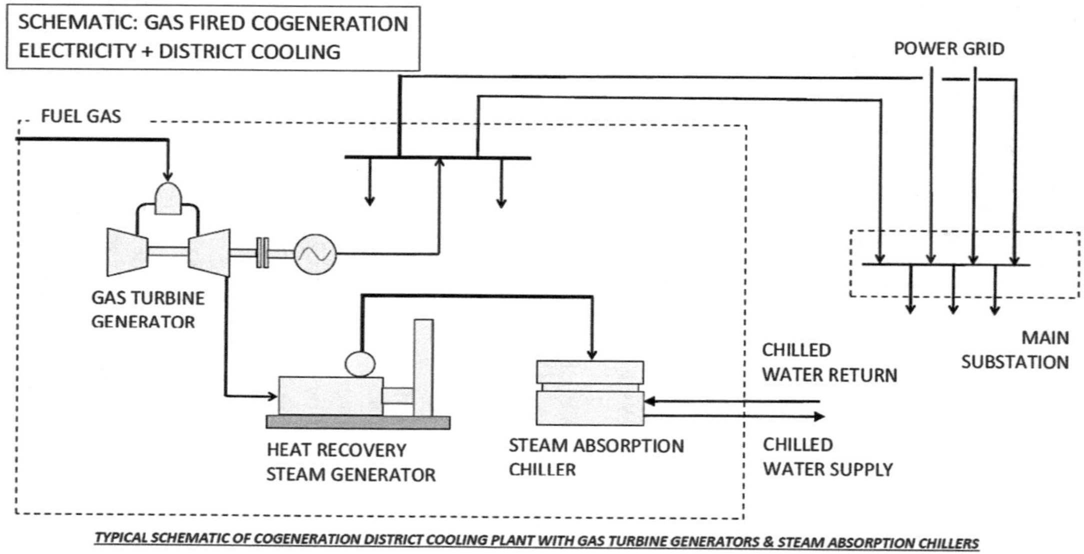

Fig. 11.1 Schematic of typical DC system layout (Source: Danny Tam Hong Khai)

Rivers are generally some degrees cooler in summer than the air, and are often used since many cities are situated on rivers. In Paris for example, a large district energy system has been gradually expanded using the Seine river both for summer cooling and (combined with other sources) for winter heating. It now provides over 400 GWh of cooling, with GHG emissions of only 35 KgCO_2e/MWh of cooling (GDF-Suez 2014; ClimEspace 2013) (Fig. 11.3).

An extremely interesting variant, seawater-based air conditioning (SWAC) fetches cold ocean water at depths of several hundred metres, using large diameter pipes with lengths up to several kilometres. Temperatures of around 6C can be obtained at 600–700 metres depth. Cool fresh water in a secondary circuit is then distributed in piping networks (Makai Ocean Engineering 2016). Many coasts have shallow continental shelves that render SWAC unfeasible; but there are a considerable number of cities with deep ocean water in proximity. This includes many oceanic islands. In northern latitudes, the sea is cold at much shallower depths, and large SWAC systems exist in major Baltic cities including Stockholm, Helsinki and Tallinn (Riipinen 2013; Hani and Koiv 2012).

Deep lakes are a good source for cold water; such systems exist in the Great Lakes region of North America and in Toronto (Enwave 2010). Lakes being fresh water offer the added advantage that the pumped water can provide city drinking water. The warmer return water on the downstream side can have useful applications for preheating, aquaculture or industry. Lakes also have stratification profiles that may offer cold water at shallow depths (Fig. 11.2).

An example is the DC system of Cornell University in the USA which supplies cooling to the campus (Peer and Joyce 2002; Cornell University 2010). Engineered by CHA Consulting/Gryphon Engineering with Hawaii-based company Makai Ocean Engineering, it has operated since 2000. Cold water at a temperature of around 4 °C is pumped from a depth of 76 m from Lake Cayuga through an HDPE intake pipe that is 1.6 m in diameter. This intake water absorbs heat through a bank of plate and frame heat exchangers from a separate closed loop that collects heat from about 100 campus buildings (8.6 million gross square feet) The DC system has a capacity of around 55MWth (16,000 tons). Efficiency is as high as 0.1 kW per ton of cooling delivered to the buildings and the annual coefficient of performance ranges from 23–29 (varying each year based on the stratified summer deep water temperature). Compared to conventional cooling, the project reduced electricity requirements by no less than 87 %, over

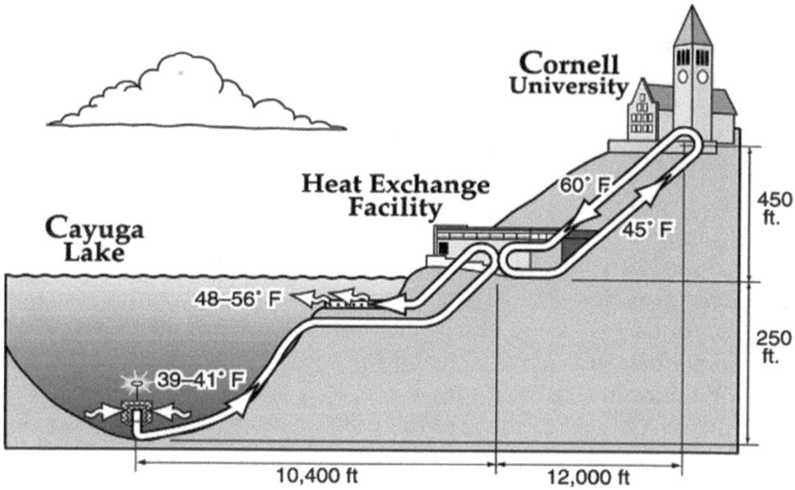

Fig. 11.2 Cornell University campus, USA: the Cayuga lake based DC (Source: Cornell University/IDEA 2010)

25 million kWh, annually and reduces peak electric demand by a total of 15 MWe. It enables annual CO_2, sulphur oxide, and NOx reductions from grid electric plants based on their mix of fuel types and eliminated the need for about 40 tons of CFC refrigerants.

The Cornell system illustrates the key considerations for good feasibility of DC: proximity of a large natural cold source; economies of scale giving high cost efficiency; supply to many buildings thus evening out the load demand; and a strong desire to reduce long-term costs. In addition, these solutions offer the bonus of having few environmental impacts. Impacts on lake or marine ecology are normally minor. They are reduced in particular by isothermal return—releasing the warmer return water to a layer of corresponding temperature in the lake or ocean. On the other hand, if a very large number of users extract energy from a relatively small river or lake, local ecosystems may be significantly affected.

Aquifers and the underground are also useful since the underground offers a temperature differential and can serve as a source of both cold and heat. It can be artificially cooled, as in central Berlin which has a large seasonal storage system. Cold water is pumped down to cool a large area underground during winter, in order to provide a source for summer

Fig. 11.3 Deep seawater intake pipes being laid out to sea for SWAC cooling at Kawaihae by Makai Ocean Engineering, Hawaii (Credit: Makai Ocean Engineering)

cooling, and similarly, heat is pumped down in summer to another part of the underground to form a heat store for the following winter (Sanner et al. 2005).

District energy solutions are, notably, omitted in the well-known McKinsey curve for reductions of global greenhouse gas emissions (McKinsey 2009). But there is now a major drive for district energy, supported by the European Union (EU Euroheat 2006), the International Energy Agency (IEA 2012), the United Nations (UNEP 2014) and industry organisations such as the International District Energy Association (IDEA 2016).

DC is found in both hot-humid climates as in Asia, and hot-dry climates such as Abu Dhabi and Dubai (Berbari 2009), where many of these are high-cost if not luxury projects, such as Jumeirah in the UAE (www.empower.ae). There has been considerable research on DC in China (Gong and Sven 2014); a proposal for a seawater combined heating and cooling system for the coastal city of Dalian (Zhen et al. 2007) concluded with high profitability for such contexts. Studies encompass both water-based and underground cooling (Song et al. 2007; The Star 2013). In the hot-humid tropics there are large systems in Hong Kong and Singapore (Lo et al. 2014; Koon et al. 2007) as well as Malaysia. Singapore has one of the most ambitious DC systems covering to date around 3 million m^2 of buildings with several hundred MW capacity (Kee 2010; Mulchand 2013). Our concept study in Chap. 8 for a large residential area in the city of Ningbo, China, includes a DC system.

Advantages, Requirements and Barriers

Interest in urban energy systems is spreading. One may add that the need for cooling is partly a result of poor planning and building design, including excessively glazed buildings. In hot climates, cooling needs can, as noted throughout this book, be reduced by better architecture, at the micro scale, as well as at the meso scale by street layouts and vegetation; but in the hot-humid tropics these 'passive' measures will often be insufficient to achieve comfort temperatures for parts of the year. Some form of mechanical cooling cannot be avoided.

The economic feasibility of DC depends on energy prices and specific urban contexts as well as issues of energy security, robustness and environment. DC already compares favourably to many conventional solutions. The large size of DC systems provides considerable economies of scale, both in technical efficiency and for the load management of a large number of customers; a DC system can reach an efficiency much higher than traditional local electricity-driven cooling. Given this factor alone, the application potential is very wide. Some systems supply customers such as a university campus or a tourist complex; this suggests that economies of scale whilst a factor do not necessarily exclude projects of a modest size. DC systems can also offer a longer life than small AC units, 25–30 years as opposed to 10 years. Typically, these large systems offer 15–40 % lower overall energy use (and hence climate emissions) than conventional AC. Where available, SWAC systems in particular offer large advantages with primary energy use reduced by up to 90 %.

Another factor for DC profitability is naturally the end-use energy density; low density suburbs require long piping distances distributing relatively little energy per kilometre. The geophysical context and city layout are also keys to feasibility, clearly best being simple street layouts in geologically easy terrain. Due to high capital costs, it is normally mandatory for all buildings within a DH/DC area to connect to the system. This implies a conflict between urban planning and building planning, discussed further later.

Although DC systems offer large savings over time, the initial capital costs are high and this is often a barrier in developing countries where cities lack financing (Fig. 11.4). But given the right conditions they are attractive for private sector investment, and can be financed and run by Energy Service Companies (ESCOs). For example, in a large development in Oslo, Norway (Arbeidsgruppe for Geotermisk Energi 2010), the private developer saw

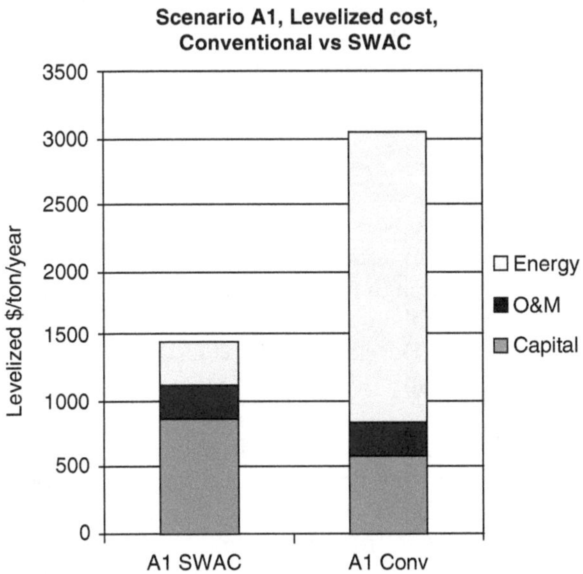

Fig. 11.4 Example, levelised cost of conventional AC versus district cooling with SWAC. Higher capital cost, much lower energy costs (O&M – operation and maintenance) (Source: Makai Ocean Engineering)

that building the energy supply based on underground heat and then running it as an ESCO offered a new business opportunity. Given rising energy prices and climate emission restrictions, the feasibility of many forms of DC is almost sure to increase rapidly in the coming years.

A notable advantage of DC systems is that they save considerable space and costs for the individual buildings, which do not need to construct or pay for expensive equipment and plant room. Customers also receive energy that is fully managed by the DC supplier and thus require very little internal energy management. A further very significant advantage is that DC offers a level of quality in operation and maintenance which is often lacking in developing countries, where AC equipment is seldom operated and maintained in an optimal manner.

In the developed economies, most cities are not expanding much; introducing DHS or DCS in existing cities involves digging up streets and reconfiguring the energy systems of connected buildings. This is very expensive, yet it is spreading fast, indicating that even at high cost, DC is becoming favourable. But DC solutions have especial relevance for developing countries where cities are expanding rapidly. In such cases DC infrastructure can be integrated from the start. Regrettably, urban planning and energy planning are seldom coordinated. Not integrating DC into these expanding hot-climate cities represents very major missed opportunities.

The Case of Malaysia

Malaysia is one of few tropical countries with significant amounts of DC, providing cooling to urban districts, airport terminals, campuses and industrial parks. However, although the Petronas Twin Towers (KLCC) in Kuala Lumpur has had DC technology for some 17 years, DC solutions are still the exception. The Asian Development Bank (ADB) reported that Malaysia could triple its DC industry (Siron and Haron 2015); this is probably an underestimate.

Malaysia has a fast-growing economy driven by the government's policy on industrialisation and commitment to becoming a developed nation by 2020. As a result, the energy sector has strong growth; in 2013, energy consumption rose by 4.6 % (Malaysia Energy Commission 2013), mainly due to increased consumption in the transportation sector. Progress includes new supply capacity as well as extending grid-connected networks in West (Peninsular) Malaysia (Basri et al. 2015; Khor and Lalchand 2014). As of 2016, total power capacity was 29,974 MW and total electricity consumption was 128,330 GWh (Malaysia Energy Statistics Commission 2016). The main energy sources are crude oil, natural gas, coal and hydro-power. Malaysia also exports gas. Renewable energy has been incentivised but is still very limited. Hence, over 90 % of primary energy in Malaysia today is oil, gas and coal (Foo 2015), indicating a strong dependency on fossil fuels for many years to come. Demand is expected to continue to rise, in line with Gross Domestic Product (GDP). According to Fazeli et al. (2016), every 1 % growth in GDP can be associated with a 1.2–1.5 % increase in energy demand; but in recent years, energy intensities have been decreasing due to energy-efficiency measures, leading to a relative decoupling of energy growth from GDP growth. At the COP15 global climate summit, Malaysia pledged 40 % reductions of CO_2 emissions by 2020 relative to the baseline year 2005. Given this difficult challenge, energy cogeneration systems are a key measure.

Cogeneration and District Cooling

Since 2015 the Energy Commission has issued 32 licenses to cogeneration plants with an overall capacity of 1066 MW; nearly 90 % of these are fuelled by natural gas, the remainder by agricultural wastes or industrial waste heat (Peninsular Malaysia Electricity Supply Industry Outlook 2016). Shaaban et al. (2011) conducted a study on opportunities and barriers to cogeneration in Malaysia. They reported that due to existing gas pipeline networks in Peninsular Malaysia, projects in urban areas are likely to be accelerated. It was also found that Malaysian energy policy with its pricing subsidies makes it necessary to maintain a good differential between gas and electricity to reflect the relative production costs. Hence cogeneration is an attractive option under the existing electricity tariff policies in Malaysia. The energy efficiency achievable with DC should become a national priority. A gas-fuelled system has many appealing features such as low capital cost, compact size, short construction time, high flexibility and reliability, fast starting and loading times, no start-up and shutdown costs, no transition costs between partial loading and full loading, lower manpower operating needs and better environmental performance (Azit and Nor 2009).

DC is often produced by cogeneration utilising the waste heat from gas turbines generating electricity for the grid. The waste heat is converted to steam and then to chilled water in a steam absorption chiller. The chilled water is distributed to cool buildings at 6 °C and returns at around 13 °C. Compared to small AC cooling units, the efficiency can be 35–40 % higher.

Many of Malaysia's DC plants also employ thermal energy storage (TES) in ice or cold water. This helps to even out energy loads, and is a particular advantage if, as in Malaysia, there are off-peak tariffs for electricity at night. The process enhances the efficiency as well as reducing waste heat to the environment. A research study estimated that one such DC system reduces greenhouse gas emissions by around 28 % (Nordin et al. 2013).

DC was first introduced in the 1990s due to the abundance of natural gas reserves in peninsular Malaysia and Sarawak; however, between 1995 and 2005 only three DC plants were developed, at Kuala Lumpur City Centre (KLCC), Kuala Lumpur International Airport (KLIA) and Putrajaya. The KLCC plant provides cooling as well as power to an area of 40 hectares, including the Petronas Twin Towers, the Kuala Lumpur Convention Centre and a large number of high-rise towers, hotels, shopping and residential complexes. This plant includes brine chillers and four units of ice storage, the brine chillers being used to charge the ice storage tanks. The pipeline distributing the chilled water network to the buildings has a 4 km loop. Another DC plant, at Universiti Teknologi Petronas in Tronoh, Perak, supplies both chilled water and electricity to the university campus. In operation since 2003, it has a capacity of 8.4 MW electric and 4000RT of cooling, and is planned to more than double to meet future demand. It is an example of a smaller, local DC plant that can be connected to the public utility for power supply in case of breakdowns or maintenance stops.

DC plants in Malaysia can be broadly categorised as follows:

1. Cogeneration DC using natural gas as primary fuel;
2. DC with cool Thermal Energy Storage (TES), powered primarily by electricity.

Cogeneration DC is made feasible by the availability of gas resources and the extensive gas piping network in Peninsular Malaysia. Cogeneration holds the promise of yielding 70 % to 80 % primary energy efficiency (Roslan and Othman 2009) together with reductions in Green House Gas emissions and other pollutant gases. This efficiency helps to reduce the energy demand thereby enhancing the nation's energy security. By generating power on-site, cogeneration also reduces the demand on the national power grid which helps to improve the resiliency of the grid. Additionally, on-site power generation avoids the transmission and distribution losses which are as high as 8.5 % (Suruhanjaya Tenaga 2017). A good example of a Cogeneration DC is the plant supplying chilled water and electricity to the

Kuala Lumpur International Airport. Completed in 1997, it has a capacity of 40 MW from gas turbine generators and a cooling capacity of up to 30,000RT using steam absorption chillers (Haron 2016). On-site power generation enhances the reliability and security of the Airport's energy system by providing it with two alternatives of power supply, the other being from the power grid.

The conventional electrically powered DC plant is usually equipped with some form of cool TES to take advantage of the peak/off-peak tariff structure offered by the power utility as part of its demand side management (DSM) strategy to encourage the shifting of air conditioning electricity consumption from peak hours to off-peak hours. For the power utility, effective DSM reduces the peak demand which in turn allows the deferment of capital expenditure on expensive peaking power plants. The strategy also cuts down on the requirement to operate inefficient peaking power plants and increases the load factor for the base load power plants, meaning higher efficiency. The DC plant serving the Kuala Lumpur International Airport 2 (KLIA-2) is a good illustration of a DC which effectively utilises cool TES (Fig. 11.5). Completed in 2012, it provides 18,000RT of cooling capacity with only 12,000RT of installed chiller capacity and 60,000RTH of chilled water TES. The use of chilled water TES in the KLIA-2 plant provides the following benefits:

1. Reduction in installed chiller capacity by 33–36 %, yielding substantial capital cost savings;
2. Reduction in utility cost by increasing chiller operation during off-peak hours, taking advantage of the lower electricity tariff;
3. Reduction in utility cost by lowering the maximum electricity demand during peak hours thus minimising the punitive maximum demand charges;
4. Improvement in energy efficiency by shifting the cooling load to night time during which the lower ambient temperature is more conducive to efficient chiller operation.

Health and comfort are important aspects of DC. The hot and humid climate makes indoor air conditioning more of a necessity for health, comfort and productivity, rather than a luxury. In Malaysian factories, typically 40 % of total power consumption is consumed for cooling, which is needed all year round.

Fig. 11.5 District cooling plant at Kuala Lumpur Airport KLIA (Source: Danny Tam Hong Khai)

The Malaysian experience also reveals obstacles that are typical in many areas of sustainable development. Building consultants, M&E contractors and facility management contractors may not be receptive to the idea of DC due to lack of familiarity with DC technology, lack of financial incentives to adopt DC technology; subscription to a DC service will inevitably reduce the building construction contract as well as the scope of building operation and maintenance (Hisham 2015). Hence these major stakeholders in the building industry may find it more to their advantage to have in-building chiller plants as compared to DC. This attests to the wide gap that often appears between good policy initiatives, and constraints in the real world.

Urban Level Versus Individual Building Solutions

There are potential conflicts between cooling individual buildings versus solutions at the level of the urban district. Which level should be prioritised? Micro, meso and macro levels involve differing interests, decision makers and stakeholders. There is a little discussed trade-off between efficient buildings and efficient large-scale supply systems such as DC. Since DC delivers a certain amount of energy per metre of expensive distribution networks, profitability is best at high city densities—and where the buildings are *not* energy efficient but have high energy demands. If buildings have low energy needs, DC profitability decreases. A general guiding principle is to prioritise DC in high density areas, versus individual and even self-sufficient off-grid energy solutions in suburbs. Consequently, where district energy is favourable there is less point in very high energy efficiency standards for the buildings. Applying both approaches in the same district will mean that the one competes with the other. This signals a key policy issue; the need to integrate the three levels of energy planning, urban planning and building design.

Hence, a little discussed consequence is that district energy systems may have the effect of dis-incentivising low energy buildings. This was illustrated whilst consulting on a waterfront development in Bergen, Norway (Butters et al. 2003–2004). The district energy company was intending to extend its network to this part of the city; at the same time, we were advising the developers to build very low energy buildings. In discussions with the energy company, the developers' message was: build your piping network right past our area, because we won't need you. With the reply: why add expensive triple glazing and extra insulation when we are bringing you cheap district energy?

In essence, this is a question of demand side versus supply side energy policies. Beyond the above simplified principle, there are naturally many complex considerations in energy planning. Local geological conditions and urban structure influence feasibility; as do the presence or not of effective energy efficiency codes. Local planning and property constraints may favour or hinder the alternatives. In the case of existing cities, a specific consideration is the age and type of the building stock. Some building types are easy to retrofit; others, especially historic buildings, are far more difficult to retrofit for energy efficiency. In such areas, a supply side option of district energy is even more preferable.

The potential conflict, or trade-offs, between individual versus urban scale can also be illustrated by cases such as the Western Harbour 'sustainable district' in Malmo, Sweden (Dalman et al. 1999). This pioneering waterfront area of some 30,000 inhabitants was acclaimed as the first in Europe to have 100 % renewable energy. In practice, what one sees is a variety of small energy solutions, chiefly solar technology on individual buildings; but over 90 % of the renewable energy is supplied by large-scale wind power located outside the city. The solar technologies pose constraints on the architecture—sometimes resulting in exciting designs, sometimes not—and are in many cases rather expensive. Hence the question: are such constraints worth it, in order to supply only a few percent of the energy? This example is not unique; some cities are now requiring individual buildings to produce a percentage of their energy with on-site renewable energy.

Most commonly this results in solar PV roof systems. Whilst this is positive seen in isolation, the 'missing policy link' is still the question of whether the energy would be supplied more efficiently and cheaply at urban scale rather than individually.

There are various design options using natural or 'passive' means to create a comfortable indoor climate, but as discussed in this book, the options are fewer in the hot-humid climates (Butters 2014); passive strategies both in city layout and individual building design can reduce but not fully eliminate space cooling needs. For cooling especially, the supply side remains important, and DC is a very high priority, not least due its unique feature of mitigating UHI in hot cities.

Issues for Urban Planning and Energy Policy

A critical issue is thus the lack of coordination between the three levels of buildings, urban plan and energy plan. Architects and building engineers have one focus, urban planners another; and few address the field of energy systems. Energy planners for their part have had little influence on urban planning. The energy policy implications of DC pertain on three levels. First, integration of urban planning and energy planning; second, integrating the levels of individual buildings and urban planning; and third, energy-efficiency requirements that are differentiated in different contexts.

The first of these involves integrating two disciplines that traditionally have had little contact; the urban planners and the energy planners. City layouts have a large influence on energy requirements. Specific layouts will be more or less favourable for district energy, in function of densities, street configurations and not least location of new urban developments near useful sources and sinks for heating and/or cooling. There are also 'industrial ecology' options whereby developments can be sited close to sources of industrial waste heat. Most commonly an urban plan is made first and the energy sector is then asked to provide the 'supply'. The energy plan must, instead, be developed at the same time as the urban plan.

Second, the shaping of cities, in particular new ones, needs to include energy considerations in the spatial layouts in order to facilitate energy efficient design of the *individual* buildings. A recognised if rare example of this is 'solar access', the earliest application of which was probably in Davis, California in the 1970s; regulating the orientation of streets and heights of buildings so that all have access to sun for solar technologies on roofs. In hot climates, streets should be oriented so that all buildings receive *minimum* sun, and can utilise prevailing local winds for passive cooling. These considerations are appearing in recent 'eco-city' design.

Third, policies for individual buildings—codes, guidelines, incentives—might, rather than applying throughout, be based on zonally differentiated requirements—analogous to regional climatic zones. Where DC is favourable, energy efficiency standards for the buildings might be lower since the supply side is particularly advantageous. The extreme energy efficiency required for *passivhaus* and similar, bears a cost penalty that may not always be justifiable in contexts where the supply side is cheaper. This applies in contexts where one has a policy choice between either enforcing very expensive retrofitting, or implementing district energy.

Cogeneration of heating, cooling and electrical power in fairly small units was pioneered by amongst others Japan. Also termed 'polygeneration', this can be very favourable especially where there is both a summer cooling and a winter heating demand, and is quite widespread in Japanese residential developments. This illustrates good integration in the planning of our three levels of buildings, neighbourhoods and energy systems.

Conclusions

District cooling systems, in common with district heating and combined DC/DH systems, offer a range of advantages including reduced primary energy use and climate emissions. They can apply to business and industrial districts as well as residential ones. They are important not least in order to ameliorate urban living qualities for low-income groups, who often live and work in poor conditions and cannot afford energy amenities such as AC. Given that much of the population is now living in urban environments, this should be given more attention in hot climate cities. At the same time, the little focused conflict between urban scale versus individual energy solutions needs to be addressed.

Energy policy should be directed more towards solutions at the district or urban scale. This applies particularly to developing countries where urban growth is very rapid. Large opportunities for sustainable development, in particular cooler cities with lower energy use and climate emissions, are being missed by DC not being integrated into new city areas.

Cities, especially in the hot climate developing countries, are rapidly growing sources of energy use and climate emissions. Efficiency improvements in buildings or in AC technology does not stop, and at best only slows, the growth of energy use and emissions. UHI should be a major factor in future energy policy decisions. In addition to reducing energy needs and climate emissions, there is an increasingly strong argument in favour of DC in order to remove the sources of heat from the city environment. District cooling offers almost the only path towards reducing the urban heat island effect. It is thus, a key to long-term improvement in deteriorating living environments in hot climate cities.

References

Arbeidsgruppe for Geotermisk Energi. (2010). *Innsatsgruppe Fornybar termisk energy*. Energi 21. Forskningsrådet (The National Research Council), Oslo.

Argonne National Laboratory. (1977). *Feasibility of a district cooling system using natural cold water*. Report for the US Department of Energy, National Technical Information Service, Springfield.

Azit, A. H., & Nor, K. M. (2009). Optimal sizing for a gas-fired grid-connected cogeneration system planning. *IEEE Transactions on Energy Conversion, 24*(4), 950–958.

Basri, N. A., et al. (2015). Malaysia energy strategy towards sustainability: A panoramic overview of the benefits and challenges. *Renewable and Sustainable Energy Reviews, 42*, 1094–1105.

Berbari, G. (2009). *History of district cooling in the Middle East*, UAE. Available at: http://www.dcproeng.com/article.php?article=144

Butters, C. (2014). *Enhancing air movement by passive means in hot climate buildings*. ELITH online publication W09, Warwick. Available at: http://www2.warwick.ac.uk/fac/sci/eng/elith/publications/

Butters, C., et al. (2003–2004). Unpublished consultancy reports for Damsgaardsundet sustainable waterfront, Bergen.

ClimEspace. (2013). *Discover district cooling and combine efficiency with urban ecology*. Available at: http://www.climespace.fr/eng/The-company/Press-room/Brochures

Cornell University/IDEA. (2010). *Case study series: College campus CHP, Cornell Lake Source Cooling System*, MA. Available at: https://www.districtenergy.org/assets/pdfs/CHP_Case_Studies/cornell.pdf, www.idea@districtenergy.org

Dalin, P. (2012). Capital cooling group. *'Free cooling'/natural cooling is crucial for a successful district cooling development*. Euroheat and Power Annual Conference, April 26–27, 2012, Copenhagen.

Dalman, E., et al. (1999). *Kvalitetsprogram for Framtidsstaden Bo01*, Malmo.

EMPOWER. Emirates Central Cooling Systems Corporation, Dubai. Available at: http://www.empower.ae/news-detail.php?id=242

Enwave. (2010). *Deep lake water cooling: Keeping Toronto cool*. Enwave District Energy, Toronto.

EU Euroheat & Power. (2006). *Possibilities with more district cooling in Europe.* Work Package 53. Brussels 2006. Available at: www.ecoheatcool.org

European Commission, RESCUE Project. (2015). *Best practice examples of district cooling systems.* Available at: https://ec.europa.eu/energy/intelligent/projects/en/projects/rescue

Fazeli, A., et al. (2016). Malaysia's stand on municipal solid waste conversion to energy: A review. *Renewable and Sustainable Energy Reviews, 58,* 1007–1016.

Foo, K. Y. (2015). A vision on the opportunities, policies and coping strategies for the energy security and green energy development in Malaysia. *Renewable and Sustainable Energy Reviews, 51,* 1477–1498.

GDF-Suez. (2014). *ClimEspace, 1st european cooling district system.* Sheet 7. Available at: www.paris-green.com

Gong, M., & Sven, W. (2014). *District heating research in China,* Svensk Fjärrvärme AB.

Hani, A., & Koiv, T.-A. (2012). The preliminary research of sea water district heating and cooling for Tallinn coastal area. *Smart Grid and Renewable Energy, 2012*(3), 246–252.

Haron, S. (2016). *Implementation of gas district cooling and cogeneration systems in Malaysia.* TNEC brochure. www.tnec.com.my

Hisham, A. (2015). *District cooling systems industry: Is regulatory framework the answer?* Available at: https://www.academia.edu/16837276/DISTRICT_COOLING_SYSTEMS_IS_REGULATORY_FRAMEWORK_THE_ANSWER

International District Energy Association, IDEA. (2016). *District Energy* (Vol. 102, no.2). www.districtenergy.org

International Energy Agency. (2012) *Implementing agreement on district heating and cooling including combined heat and power.* http://www.iea-dhc.org/the-technology.html

Kee, T. P. (2010, June). *District cooling as an energy and economically efficient urban utility—Its implementation at Marina Bay business district in Singapore.* Singapore District Cooling Pte Ltd.

Khor, C. S., & Lalchand, G. (2014). A review on sustainable power generation in Malaysia to 2030: Historical perspective, current assessment, and future strategies. *Renewable and Sustainable Energy Reviews, 29,* 952–960.

Koon, C. H., Sim, J. S., Ng, Y. S., Tay, L. C., & Ng, C. B. (2007). *Indirect seawater cooling and thermal storage system in Changi Naval Base.* Singapore.

Lo, A., Lau, B., Cheng, V., & Cheung, P. (2014). *Challenges of district cooling system (DCS) implementation in Hong Kong.* Arup Hong Kong, SB14, Barcelona.

Makai Ocean Engineering. (2016). www.makai.com. [Our thanks to Dale Jensen, Makai, Hawaii for information, comments and images provided].

Malaysia Energy Commission (2013). *Suruhanjaya Tenaga (2013)* [*National Energy Balance 2013*].

McKinsey & Company. (2009). *Pathways to a low carbon economy*. Version 2 of the global greenhouse gas abatement cost curve. https://www.cbd.int/financial/doc/Pathwaystoalowcarboneconomy.pdf

Mulchand, A. (2013, November 24). S'pore has world's largest district cooling plant. *The Straits Times*, Singapore. http://news.asiaone.com/news/singapore/spore-has-worlds-largest-district-cooling-plant

Munck, C., Pigeon, G., Masson, V., Meunier, F., et al. (2013). How much can air conditioning increase air temperature for a city like Paris, France? *International Journal of Climatology, 33*, 201–227.

Nordin, A., Norsheila Buyamin, M., Majid, A. A., Amear, S., & Ariffin, S. (2013, June). Evaluation of carbon dioxide emission using energy analysis approach: A case study of a district cooling plant. *International Journal of Computer and Electrical Engineering, 5*(3), 284–287.

Peer, T., & Joyce, W. S. (2002, April). Lake-source cooling. *ASHRAE Journal, 44*, 37–39.

Peninsular Malaysia Electricity Supply Industry Outlook. (2016). http://www.st.gov.my/index.php/en/component/k2/item/680-peninsular-malaysiaelectricity-supply-industry-outlook-2016-download.html.

Radspieler, A., Xu, P., & Haves, P. (1999). *Seawater source cooling for air conditioning commercial buildings*. California Energy Commission, PIER Program CEC-500-99-013, Lawrence Berkeley National Laboratory.

Riipinen, M., (2013). *Modern district heating and cooling systems*. Clean Energy Ministerial CHP/DHC Working Group, International Energy Agency CHP/DHC Collaborative Joint Workshop, November 26–27, 2013, Helsingin Energia.

Roslan, M. F., & Othman, M. R. (2009, October). *Sharing on gas district cooling cooling (GDC) in Malaysia*. www.igu.org. In 24th World Gas Conference.

Sanner, B., Kabus, F., Seibt, P., & Bartels, J. (2005). *Underground thermal energy storage for the German Parliament in Berlin, system concept and operational experiences*. In Proceedings of World Geothermal Congress 2005, Antalya, April 24–29, 2005.

Shaaban, M., et al. (2011). Grid integration policies of gas-fired cogeneration in Peninsular Malaysia: Fallacies and counterexamples. *Energy Policy, 39*(9), 5063–5075.

Siron, M. H. A., & Haron, M. H. (2015). *Energy and economic assessment of district cooling system in Malaysia*. In Science & Engineering Technology National Conference, Bangi-Putrajaya, Selangor.

Song, Y. H., Yasunori, A., & Jung-Jae, Y. (2007). Effects of utilizing seawater as a cooling source system in a commercial complex. *Energy and Buildings, 39*, 1080–1087.

State of Hawaii. (2002, October). *Sea water district cooling feasibility analysis for the state of Hawaii.* Department of Business, Economic Development & Tourism's (DBEDT) Energy, Resources, and Technology Division, Hawaii.

Stewart, I., & Kennedy, C. (2015). *Estimating anthropogenic heat release from megacities.* In ICUC9 – 9th International Conference on Urban Climate held jointly with the 12th Symposium on the Urban Environment, July 20–24, 2015, Toulouse, France University of Toronto, Toronto.

Suruhanjaya Tenaga. (2017). Review on electricity tariff in Peninsular Malaysia under the incentive-based regulation mechanism (FY2014-FY2017). Available at: www.st.gov.my

Suruhanjaya Tenaga (Malaysia Energy Statistics Commission). (2016). Malaysia Energy Statistics Handbook 2016, Putrajaya.

The Star. (2013). Available at: http://www.thestar.com.my/business/business-news/2013/10/23/ and South East Asia Energy Efficiency Project, https://www.adb.org/projects/46914-014/main

TNEC. TNB Engineering Corporation Sdn Bhd, brochure. Available at: www.tnec.com.my

UNEP. (2014). *District energy in cities: Unlocking the potential of energy efficiency and renewable energy.* Paris. http://www.unep.org/energy/DistrictEnergyinCities/tabid/816789/language/enUS/Default.aspx

United Nations, Department of Economic and Social Affairs, Population Division. (2014). *World urbanization prospects: The 2014 revision.* UN: New York.

Zhen, L., Lin, D. M., Shu, H. W., Shuang, J., & Zhu, Y. X. (2007). District cooling and heating with seawater as heat source and sink in Dalian, China. *Renewable Energy, 32*(15), 2603–2616.

CHAPTER 12

Cooler Cities: What Kinds of City?

Urban Form, Climate and Wellbeing

Chris Butters

Introduction

How can we ensure cooler cities, in hot developing countries in particular? Our focus has been on Asia, home to most of the fastest growing cities in the world, many of them in extremely demanding tropical climates. Cities have been formed in many ways through the course of history. Perhaps we need new city paradigms for the future; or can some tried and trusted solutions, given modern methods, perhaps provide the best answers? Do some conclusions emerge about what type of cities we should choose to build?

In former times, towns did develop on the basis of simple environmental factors, such as favourable climatic sites, natural resources and availability of water, in addition to factors including defence and proximity to trade routes. But the basic physical, environmental parameters, such as location, terrain, climate, energy or water have been largely secondary considerations in the growth of our recent cities. Major city planning decisions in the real world are seldom made with environmental grounds as a high priority, but on a wealth of pragmatic considerations such as land availability, land prices, existing infrastructures and market interests, or on grounds related to

C. Butters (✉)
University of Warwick, Coventry, UK

A. Cheshmehzangi, C. Butters (eds.), *Designing Cooler Cities*,
Palgrave Series in Asia and Pacific Studies,
https://doi.org/10.1007/978-981-10-6638-2_12

national and regional policies, prestige and much else. However, we do see that environmental considerations are gaining in focus, as witnessed by the efforts of hot climate cities to improve their microclimate—even to the extent of restricting development densities, that is to say the ‘bottom line’ of profit. One might see in this a shift back towards a more centrally planned approach to planning, which has been out of fashion in neo-liberal economies; but with the rationale now being partly environmental rather than economic or ideological. Nevertheless, we can approach the question of urban form from a principally environmental point of view. Seen in purely physical terms, what kinds of city can provide the coolest environments in hot climates? In this chapter we discuss some of the environmental arguments for and against different city paradigms.

Several ideals and paradigms for cities coexist today. These include dense high-rise cities, compact cities, traditional European-type cities (which are found all over the globe), garden cities and low-dense cities. These are variously championed as being advantageous in terms of economy, social qualities, environment and more (Jabareen 2006). It goes without saying that cultural traditions and preferences play a large role too. More recently, ‘eco-city’ visions and projects around the world have emerged, which broadly follow one or other of these paradigms, or indeed mixed elements of several. In the large cities of the hot-climate developing countries, and perhaps especially in Asia, the paradigm of dense high-rise development is certainly predominant.

We have argued that creating cooler cities in hot climates, and mitigating their energy footprint and climate emissions, requires good solutions on all three levels of individual buildings, urban layouts and overall energy supply. This demands that we link engineering and architecture with landscape design, urban planning and energy planning. Although we may identify the best solutions on each level, at least equally important is to determine which level it is most sensible to prioritise in varied contexts, and the interactions between levels. This renders a ‘science’ of eco-friendly form and design much more complex. For example, in the specialised field of ventilation and air movement for design of cooler buildings or streets, our co-authors have noted the many variables that even advanced computer models and CFD (computational fluid dynamics) struggle to handle. There are hundreds of studies in the field of passive cooling design, but it is difficult to provide a comparative basis of *which combinations* of solutions are best in a given context, or *on which level* of micro, meso or macro it is best to apply our efforts. To add to the complexity of the task, that word ‘best’ must also include cost and social considerations. Nevertheless, in this chapter we note some of the major environmental and microclimatic considerations regarding our choice of city paradigms, and where these appear to offer significant advantages or disadvantages for cooler cities.

Fig. 12.1 Urban green infrastructures are important parts of city environments to combat the UHI effect. They provide breathing spaces: places to socialise, relax and do various exercises. Here we see a good combination of blue/green spaces, local people socialising in their urban community park, and visitors enjoying the scenery in a hot climate. ZhongShan Park of Ningbo City, China (Source: Ali Cheshmehzangi).

Location

This first key to cooler cities hardly needs further elaboration; general considerations for cities in hot climates include choosing convex locations rather than concave ones (heat traps), sites facing prevailing breezes, proximity to water bodies, and green surrounds. Location may make a difference to the city environment of several degrees. Due to land pressures, existing cities often densify or expand into climatically unfavourable areas, such as poorly ventilated valleys or polluted zones. This is even more the case with low-income and informal settlements. Traditional settlements often chose very favourable locations; but climatic location is hardly considered today when cities are choosing where to expand. In practice, there is not often much choice; since new areas of urbanisation usually form an expansion of existing cities, where existing infrastructures and other pragmatic factors are paramount.

Layout

The meso level of city planning and layout has been addressed in several contributions in this book. Overall layout should be fairly compact so as to facilitate public transports (as well as walkability); street layouts and orientation should enable all buildings to minimise incident solar radiation and to use local breezes for maximum urban ventilation. The urban structure should also be designed with an economical layout for district scale energy as well as the other infrastructures, such as water supply and sanitation. Furthermore, other factors such as building volumes, heights and setbacks should be determined so as to facilitate passive cooling design, good natural daylighting and maximum cross-ventilation.

As we return to later, not only the streets and buildings but the open spaces and green-blue infrastructures also play a large role in city temperatures as well as living quality, and these elements must be seen together. When well-located and designed within the overall layout, the urban parks and green spaces can lower the temperature in surrounding city blocks considerably. In common with the above, these are evident considerations, but encounter many barriers in the real world, due to factors such as existing property boundaries and street layouts as well as profit maximisation by developers.

Fig. 12.2 China's remarkable Tulou houses represent some of the best examples of vernacular architecture. These structures represent several dimensions of sustainability: social (by creating strong local communities), environmental (by using local materials and enhancing passive cooling), and economic (by durable construction methods with low costs and maintenance). Fujian Province, China (Source: Chris Butters).

Density

This third point is central to our ideas of how we think cities should be. The ideal in many of the fast-expanding economies of Asia today, as in the Persian Gulf States and other world regions, is still a rather uncritical trend of high-rise concepts and an outdated zoning approach from the modernist era. In addition, comes the free rein still given to private car transport with its effect of heating up the city. Naturally, the concept of 'compact cities' has a range of interpretations (Burgess 2000). Dense, high-rise cities offer some advantages in terms of sustainability, but have considerable downsides too. Extreme density certainly increases congestion and

Table 12.1 Urban density comparisons

Urban typology	*SC—Surface coverage*	*FAR – Floor area ratio*	*Average height*
1. Europe, detached housing	0.10–0.30	0.2–0.7	1.5–2.5
2. Europe, row/terrace housing	0.15–0.35	0.5–1.0	2.0–3.0
3. Ningbo low-rise traditional	0.50	1.4	2.4
4. Ningbo 6-storey block 1990s	0.23	1.2	5.0
5. Jinan low blocks 1980s	0.34	1.8	5.3
6. Europe modernist high-rise	0.10–0.25	1.0–2.5	8.0–14.0
7. Jinan superblock 1990s	0.22	2.0	10.1
8. Ningbo high-rise block	0.17	2.6	15.5
9. Europe compact city block	0.35–0.55	2.0–4.0	4.0–6.0

Notes:
1. Average height (number of floors) is less than the high-rises themselves due to some low-rise commercial and other buildings on most sites. For example, the high-rises themselves in example 8 are around 30 stories
2. Figures are indicative. Sources: Ningbo cases (Cheshmehzangi and Butters 2015), Jinan cases (Yang 2010), Europe cases (LSE Cities/Eifer 2014)

pollution, as well as aggravated heat island effects. An evident basic fact is that the denser the city, the more heat it produces in a given area.

A key reasoning behind dense, high-rise development is often thought to be the need to house large populations in a small area. This argument is not strictly true; high population densities are achievable with quite low-rise cities. Table 12.1 provides density comparisons from various sources showing typical figures of Surface Coverage (SC) and Floor Area Ratio (FAR), two principal measures of urban density (see Chap. 8), as well as building heights in number of floors. Note that these are densities within the site limits themselves, excluding adjacent areas of roads, parks or other spaces. Typical FAR figures range from below 1.0 in suburbs to above 4.0 in some inner-city areas. The FAR in quite dense low-rise districts seldom exceeds 2.0. FAR above around 2.5 can be considered as high-density. It is notable that inner city areas in many typical European cities have a FAR of over 3.0 and even 4.0. These traditional cities actually have *a higher population density* than the high-rise 'superblocks' in many Chinese and similar new cities.

As illustrated, there are, thus, various city types that achieve high population density. It may also be noted that overall city population densities—including, the roads, green spaces and so on are not very different in old

European cities and in modern Chinese or other new high-rise cities; most of the highest overall population densities are to be found in crowded low-rise Indian and similar cities (Demographia 2016). Hence, there is little reason in terms of population pressure to adopt dense high-rise models.

Low-Rise and High-Rise

As regards height, measures and perceptions vary too. The term low-rise is used here to mean up to about six floors (possibly up to around eight floors in parts of an area). This corresponds to typical traditional European cities, as well as being chosen in quite many recent 'sustainable city' projects. The term high-rise, thus, refers here to heights ranging from around eight floors to extremely high-rise solutions of 30 and more floors, which are now not uncommon in the world's growing megacities. Common classifications also often include mid-rise typology, which we do not use here. These definitions vary quite widely from country to country.

In hot climates, any form of high density aggravates UHI, thus increasing the need for cooling; but high-rise brings specific climatic disadvantages. They are for cost reasons normally quite deep buildings, leading to apartments with one-sided ventilation and poor daylighting, hence poor comfort as well as greatly increased energy use for lighting and mechanical cooling. In low-density areas by contrast, there is often sufficient—and free—natural ventilation. In large high-rise buildings, many apartments or offices are also of necessity unfavourably oriented, facing for example only towards troublesome east or west sun, again increasing cooling needs.

Furthermore, in high-rises, many of the well-recognised passive solutions for cooling are not possible. Solar protection, a key to cooling in hot climates, is difficult since much more of high-rise facades are exposed (Jianlei 2004). High-rise buildings cannot be shaded by trees, whereas low buildings can be; one cannot use courtyards and similar vernacular solutions for passive cooling; lightweight materials, favourable in hot-humid climates, are less feasible in high-rise.

In addition, detailed studies such as those in Table 12.1 show that high-rise urban typologies are, in terms of thermal energy performance, not better than low-rise. Operational energy efficiency can be just as good in low-rise; and generally better than what is achievable in very dense high-rise areas.

Fig. 12.3 'Informal settlements' or slums are almost universal in hot climate developing countries. Usually characterised by overcrowding, haphazard design and few comfort amenities or green spaces, their inhabitants also often suffer the most from urban heat events. Informal settlements in Bangalore, India (Source: Ali Cheshmehzangi).

Green Space

Achieving a good urban microclimate is a key factor for reducing energy use and cooling needs. UHI, being a serious issue in hot climates, is also very much an issue of public health and comfort (Sakka et al. 2012). A city's green-blue infrastructures—trees, parks and water bodies in particular—contribute greatly to cooling the built environments, as well as filtering noise and pollution and providing wellbeing (Zhang and Wang 2015). City temperatures can be several degrees lower in the vicinity of parks (Yu et al. 2006). Many cities are making major efforts in this field.

There is a key difference between typical new high-rise areas and traditional European-type cities. The former have considerable green space between the high-rises, but these are often privatised and little used areas; the European city blocks have only small green spaces locally but large urban parks nearby. In addition to being truly public and accessible to all, these parks are often also large, enabling higher biodiversity, more activities and socialising than the often gated spaces between high-rise blocks. It may be added that in dense inner cities, the many urban infrastructures also represent major land use interventions on almost the whole area, leaving little room for undisturbed natural areas. In high-rise city developments, the seemingly 'green' areas between buildings often consist only of a thin green layer on top of extensive engineering works, such as underground parking and infrastructural engineering works.

Cooling also has an important equity aspect. It is the poorest groups who often suffer most the effects of high urban temperatures, since they must more commonly commute and often work outdoors, in addition to lacking cooling amenities in their homes and workplaces. Whilst business districts can at least afford to pay for cooling, UHI is most disadvantageous for low-income contexts. This is the situation for millions in the expanding cities of the emerging economies. Hence, the outdoor city environment is as important as the indoor solutions for cooling.

Fig. 12.4 Grey versus Green is a major debate in urban planning. Phnom Penh city in Cambodia is an extreme case of the trend of disappearing green and blue areas in cities, where new housing and commercial projects are replacing existing lakes, water corridors and open green spaces. The figure is from the waterfront park which is almost the last remaining publicly accessible green space in Phnom Penh, Cambodia (Source: Ali Cheshmehzangi).

Energy and Renewables in the City

As noted in Chap. 11 on District Cooling, the fundamental challenge for cooler cities is to remove the sources of unwanted heat from the city. This applies especially to vehicles and to air conditioning (AC) in buildings, which are the two major anthropogenic sources of city heat. More efficient cooling technology, including heat recovery systems, is a part-solution; however, the rapid spread of AC, that is to say the sheer increase in volume of cooling, outweighs the efficiency gains in many cases.

Buildings have many other needs for energy including for lighting, cooking and appliances. It is broadly accepted policy that future buildings should where possible produce their own energy with renewable sources—solar in particular. This raises more questions as to the type of city.

Renewable energy supplies (RES) integrated into roofs and facades in the form of both thermal and photovoltaic (PV) solar systems are spreading, and can cover the entire energy demand when it is as small as in *passivhaus* type buildings. 'Plus-energy' buildings that produce more energy than they need are already well known (Disch 2017). In contrast to AC, these do not significantly produce unwanted waste heat to the urban environment. In this way, a town can become its own 'power station'. But this is *only* possible in low-rise typologies of up to around four floors. In tropical latitudes where the solar angle is high, solar panels must be close to horizontal to achieve good efficiency; in other words, primarily on rooftop surfaces. A PV roof on a skyscraper might be enough to power the lifts and little more. In dense high-rise areas, the tall buildings also tend to shade each other more than in low-rise, limiting 'solar access' for neighbouring buildings. Compact city forms are, on the other hand, advantageous for district cooling systems, which are an absolutely key solution wherever they can be implemented. However, as we have noted these presuppose financial and governance capacity that is not available in many cities in less developed countries.

Some futuristic proposals have included wind turbines on buildings. This is little more than symbolic 'eco-bling' (Liddell 2013), and does not make much sense for several reasons, including noise and vibrations, in addition to the fact that relatively small wind turbines produce little energy per swept area and that the inner-city location is very seldom where available wind speeds are high. On the other hand, large industries, as well as some urban waste streams such as bio-effluents and solid waste contain large quantities of waste energy that can be converted to cooling; but this again normally happens outside the city limits, thus not adding to UHI. Natural ventilation cooling by stack effect is in principle advantageous in tall buildings since the height provides greater air pressure differences. However, the generally polluted city air in dense inner cities makes this option undesirable. In extreme cases, already today, the outdoor air in cities is so polluted that far from wanting natural air flow indoors, one needs to filter the outdoor air or keep it out altogether.

In summary, low-rise types of cities are advantageous for renewable energy production; as well as for application of free, passive cooling techniques, such as solar chimneys and evaporative cooling. These should be priorities at least in the parts of a city that can be fairly low-rise.

Transports

Vehicle use typically accounts for around one-third of the total energy use (and unwanted heat production) in cities (Steemers 2003). Whilst we do not focus on transport, it is evident that cooler cities require both reducing traffic by better public mobility systems, as well as solutions such as electric vehicles where the production of the electricity with its associated waste heat is, again, moved outside the city itself. In order to reduce urban heat from transport—pollution being obviously just as important—this also involves our choices of what type of city. Well-known planning approaches include transport-oriented development (TOD), which aims to minimise needs for movement of citizens, goods and services by shaping the city in efficient ways. Further, the key eco-city principle of mixed use co-locates homes, workplaces and services in order to reduce the needs for travel. These are strategies on the meso level of city layout planning.

Mobility is a main sustainability issue for energy, economy and society. Megacities require extensive and extremely costly multi-level transport solutions (underground, surface and elevated). Despite these, vehicle speed is often little above walking speed; congestion multiplies heat emissions and pollution (Baker and Steemers 2000). At the other extreme we have suburban sprawl, which is extremely inefficient in terms of transports. Between those extremes: 'Interestingly, historic European cities, such as Paris, lie at a national "optimum"—achieving moderate energy use for modest densities—whilst sustaining a rich urban life. It is not evident that moving towards increasing urban density will lead to reduced car traffic—in fact, in the short-term the opposite is likely to be the case. In the absence of extra capacity in the form of effective integrated public transport, increasing the density will inevitably increase traffic' (Steemers 2003). This, again, suggests how densification is often not positive in planning practice.

Proposals as in some eco-city concepts for transport on multiple levels, separating all vehicle traffic from pedestrian areas—often illustrated with lots of attractive vegetation and trees on the upper pedestrian levels—are far from new. Apart from metro systems, unavoidable tunnels and occasional 'sky gardens', they are seldom implemented to any extent for the simple reason that this is prohibitively expensive. The costs and environmental effects are evident in a few cities that have had to resort to elevated multi-level highways in the inner city, such as Bangkok.

A car-oriented city may have somewhat lower overall population density due to the large areas of streets and high *vehicle* density—at a high price of pollution, noise and fuel use. An advantage of a more 'car-free' city is that one can achieve higher population densities, as well as environmental quality. This also carries a business argument: in a car-free district, a higher proportion of the extremely expensive urban land can be capitalised instead of devoted to cars or parking. But there are conditions for a city to achieve lower vehicle use: it must have mixed-use zoning (homes, work, services and leisure not separated by long distances), it must be fairly small (short walking distances) and have good public transport facilities. In cities such as Freiburg or Oslo, people do not need cars for most activities, and many do not have cars at all (Nielsen et al. 2007). To achieve compact mixed-use, a city cannot be a 'car city'. In addition to the above, roadways, parking spaces and transport areas take up a very large part of city land and these are hard surfaces that absorb solar radiation and greatly increase UHI.

Fig. 12.5 The Marina Bay project in Singapore is one of the largest district cooling projects in the world. It illustrates integration of sustainable urban energy systems into large urban design and architectural projects. Marina Sands Bay, Singapore (Source: Ali Cheshmehzangi).

Table 12.2 Embodied carbon of typical modern buildings

Embodied carbon: materials		
In many buildings, concrete and steel comprise by far the major part of the total carbon footprint		
Concretes + steel:		
Sweden, 4-storey offices	81 %	Source: Wallhagen et al. (2011)
Italy, 6-storey apartment block	76 %	Source: Our analysis, from Blenghini (2009)
China, high-rise office building	>70 %	Source: Zhang and Wang (2015)

Embodied Energy/Carbon

Embodied energy (EE) or carbon is, as noted in our introduction, a large and growing component of overall energy use in the built environment (Ibn-Mohammed et al. 2013; Sartori and Hestnes 2007). This EE does not produce unwanted urban heat since almost all materials are produced outside the city, except for the heat produced on construction sites and by trucks bringing materials, and in maintenance (recurrent energy inputs). But EE does comprise large climate emissions, which contribute to GW; and thus indirectly worsen city climates. This again is influenced by *what type of city* we choose to build, and so does merit a brief word too.

To reduce EE with its concomitant contribution to climate emissions and global warming, we must replace energy-intensive construction materials. As shown in Table 12.2, many studies show that the largest EE items in a building LCA (life cycle analysis) are often cement products and steel; these two alone often comprise 70–90 % of the total EE. Other energy-intensive materials include various metals, glass and plastics. EE will almost inevitably be higher in dense, compact and high-rise cities due to the need for energy-intensive materials, such as concrete, steel, glass, aluminium, fire protective materials and so on. This is a disadvantage of compact high-rise types of city, which many may see as minor; but the part of EE in the total picture will become *significantly greater* in coming years since the operational energy required to run and cool buildings is a much smaller part in tomorrow's low-energy buildings.

It has been little noted that the above applies not only to city buildings but also to the vast urban infrastructures—flyovers, subways, underground parking areas, piping, culverts, and so on—necessary in dense cities. These almost inescapably have to be in energy-intensive materials, such as concrete and steel. We have noted elsewhere (Butters and Thomas 2017) that in a high-rise type city development, these may embody half as much again of

EE as all the building works above ground—some one-third of the total. The site works and infrastructures are a little perceived part of the picture. Again, this relates to our choice of city paradigm.

Another common argument in favour of compact cities is economies of scale. But the presumed economies of scale in high-density infrastructures have been seriously questioned: 'Other empirical studies have consistently found that lower operating costs in the suburbs more than offset the higher initial capital costs of installing new infrastructure' (O'Toole 1996). Similarly, it is also shown that 'the cost increase from the low to the medium-density scenario is nearly 100 %, while the cost increase from the medium to the high-density scenario is just over 50 %' (Barter 2000). These are strong arguments against the common belief in high density and economies of scale.

Other Paradigms

The 'Garden City' paradigm, initiated by Ebenezer Howard (1902) as a response to miserable conditions in nineteenth-century industrial cities—not unlike conditions in parts of today's Asian cities—was conceived to provide comfort and health for working classes by bringing fresh air and green spaces into the city combined with fairly low building densities. This concept has found renewed interest in recent years. The example from India in Chap. 7 illustrates such a residential neighbourhood. The FAR in garden city developments may reach up to around 1.3; whilst therefore not appropriate where high population densities are unavoidable, it remains a valid approach for smaller towns and allows for 'design with nature' including the full range of passive design measures to maximise natural cooling.

'Eco-city' goals include all three areas of environmental, economic and social sustainability, such as energy efficiency, low carbon footprint, mixed use, walkability, public health, economic and social diversity. There are also important 'new' conversations concerning urban governance, resilience and sustainable consumption. It is notable that many of the most successful urban eco-districts of recent years in Europe have been low-dense developments. In these, the FAR may be around 2.0 as a maximum, with building heights of typically 5–6 floors and quite plentiful green space. This is partly made possible by mixed use along with excellent public transport and sustainable mobility solutions that really reduce traffic, as well as reducing roadway areas. The Vauban District in Freiburg, Germany is a leading example with around 20 years of experience (Butters et al. 2010). The social qualities of such districts are a main attraction. With this type of low-rise paradigm, it is also possible for the buildings themselves to provide their own energy with PV roofs, as well as applying all available passive design techniques—whether for heating as in the German context or for cooling in a hot climate context.

Amongst the more futuristic concepts for eco-cities, many rely on advanced technology, rather than natural means, to provide cool conditions. One must retain some scepticism as to whether these visions really offer 'sustainable cities', since they may have an impressive array of eco-technical hardware but often lack a real economic basis or a sustainable social community. Multi-level transport has been mentioned; another is the sheer cost of inappropriate locations, such as deserts, both for economic activity and for basic needs such as food production and fresh water. The required eco-technology is available but massive, and too prohibitively expensive for such visions to be a model for developing countries.

Whilst dense high-rise may be appropriate for Central Business Districts (CBDs), the UHI effect plus energy use, as well as health and social issues are arguments that weigh somewhat against dense, high-rise city models; although naturally, there are also other, positive factors in the picture. In addressing developing-country contexts, it should be recalled that low-income urban areas tend to be especially disadvantaged. We have discussed elsewhere the equity implications of dense cities with rising UHI effects (Thomas and Butters 2017). In developing countries, where resources are often limited, dense high-rise city environments planned and built to a high level of quality may be comfortable, but *mediocre quality* high-rise ones may provide little better than 'vertical slums'.

Between the extremes of compact high-rise and suburban sprawl, however, there lies a range of options for hot climates that, given good solutions on all three of our levels, can all offer resilient, efficient, sociable and comfortably cool cities.

Conclusions

In addition to the question of cooling, we must remember the broad energy considerations related to our choices of city form; since any urban form that leads to higher energy needs is almost inevitably a hotter city, due to the waste heat associated with energy use. Besides considerations related to environment and cooling, there is evidence suggesting that moderate size, low-dense cities offer reduced complexity, lower costs, resilience and a range of social advantages compared to large, compact high-rise cities.

In an ideal world, urban policy aims for a good balance between ecological, economic and social criteria. In the real world, however, urban planning and policy decisions are normally far more heavily influenced by financial, political and pragmatic factors. What kinds of city offer the best opportunities for cooling? Dense high-rise as compared to low-rise is problematic in relation to urban heat island, pollution and other impacts of high energy density. Perhaps the advantages of low-rise, and somewhat lower density, outweigh the arguments for dense high-rise. This would seem to be the case except where dense high-rise can be of a high quality and with sustainable design on all three of our levels. Especially important for cooler cities are sustainable, non-UHI producing energy supply systems, such as district cooling. However, in many cities of developing countries, it will be a long time before such solutions are achievable. These require a multidimensional understanding of development processes, policy frameworks, regulations and, not least, political will. The passive-design solutions on micro and meso levels are both simplest and cheapest and should be a natural priority if resources are limited. In the real world of hot-climate developing-country cities, the task of creating cooler cities is still a big ask.

References

Baker, N., & Steemers, K. (2000). *Energy and environment in architecture: A technical design guide*. London: E&FN Spon.

Barter, P. A. (2000). Transport dilemmas in dense urban areas: Examples from Eastern Asia. In M. Jenks & R. Burgess (Eds.), *Compact cities, sustainable urban form for developing countries*. London/New York: Spon Press.

Blenghini, G. (2009). Life cycle of buildings, demolition and recycling potential: A case study in Turin, Italy. *Building and Environment, 44*, 319–330.

Burgess, R. (2000). The compact city debate: A global perspective. In M. Jenks & R. Burgess (Eds.), *Compact cities, sustainable urban form for developing countries*. London/New York: Spon Press.

Butters, C., & Thomas, T. H. (2017). *Comparing the carbon impact of buildings to that of nfrastructures*, ELITH online publication W07. Available at http://www2.warwick.ac.uk/fac/sci/eng/elith/publications/. Warwick University, UK.

Butters, C. et al., (Eds.). (2010). *Signals: Local action – Success stories in sustainability*. Stiftelsen Idebanken (The Ideas Bank Foundation), Oslo. See also www.vauban.de

Cheshmehzangi, A., & Butters, C. (2015). *Refining the complex urban: The study of urban residential typologies for reduced future energy and climate impacts*. In:

Proceedings of 8th conference of the international forum on urbanism, Incheon, South Korea.
Demographia. (2016). *World Urban Areas.* 12th Annual Edition April 2016. demographia.com/db-worldua-index.htm
Disch, R. (2017). Rolf Disch architects. *Plusenergiehaus—das Wohnmodul.* www.plusenergiehaus.de, http://www.rolfdisch.de/index.php?p=home&pid=78&L=0&host=2#a285 (pdf and other articles, in German, see also www.rolfdisch.de). Accessed 02 June 2017.
Ebenezer, H. (1902). *Garden cities of tomorrow*. London: S. Sonnenschein.
Ibn-Mohammed, T., Greenough, R., Taylor, S., Ozawa-Meida, L., & Acquaye, A. (2013). Operational vs. embodied emissions in buildings – A review of current trends. *Energy and Buildings, 66*, 232–245.
Jabareen, Y. R. (2006). Sustainable urban forms, their typologies, models, and concepts. *Journal of Planning Education and Research, 26*, 38–52. Association of Collegiate Schools of Planning; MIT, USA.
Jianlei, N. (2004). Some significant environmental issues in high-rise residential building design in urban areas. *Energy and Buildings, 36*, 1259–1263.
Liddell, H. (2013). *Ecominimalism: The antidote to eco-bling* (2nd ed.). London: RIBA Publishing.
LSE Cities/EIFER. (2014). *Cities and energy. Urban morphology and heat energy demand.* Final report, London.
Nielsen, G., et al. (2007). *Miljobyen Freiburg*. Oslo: Transportokonomisk Institut.
O'Toole, R. (1996). *The vanishing automobile and other urban myths.* Different Drummer, The Thoreau Institute, USA.
Sakka, A., Santamouris, M., Livada, I., Nicol, F., & Wilson, M. (2012). On the thermal performance of low income housing during heat waves. *Energy and Buildings, 49*, 69–77.
Sartori, I., & Hestnes, A. G. (2007). Energy use in the life cycle of conventional and low energy buildings: A review article. *Energy and Buildings, 39*(3), 249–257.
Steemers, K. (2003). Energy and the city: Density, buildings and transport. *Energy and Buildings, 35*(1), 3–14.
Thomas, T. H., & Butters, C. (2017). (forthcoming) Equity and thermal public health in tropical cities. *Journal of Municipal Engineering UK.*
Wallhagen, W., Glaumann, M., & Malmqvist, T. (2011). Basic building life cycle calculations to decrease contribution to climate change – Case study on an office building in Sweden. *Building and Environment, 46*, 1863–1871.
Yang, J. (2010). *Does energy follow urban form? An examination of neighborhoods and transport energy use in Jinan, China.* Cambridge, Massachusetts, USA: MIT.
Yu, C., Chen, Y., & Nyuk Hien, W. (2006). Thermal benefits of city parks. *Energy and Buildings, 38*, 105–120.
Zhang, X., & Wang, F. (2015). Life-cycle assessment and control measures for carbon emissions of typical buildings in China. *Building and Environment, 86*, 89–97.

CHAPTER 13

Conclusions and Recommendations

Ali Cheshmehzangi and Chris Butters

A Brief Summary

We have argued that solutions for cooler cities in hot climates must be addressed on all three of the levels discussed in this book. Urban cooling, with its associated high energy use and climate emissions, is influenced by building design, city form and green infrastructures, and overall urban energy systems. Choice of low-dense or high-rise is a key factor, which in addition impacts on embodied energy and carbon. Urban density and structure also have a large influence on transport needs, not discussed here; and all the above aggravate the growing urban heat island effects in many parts of the world. Our Asian cases present some contemporary research and innovative strategies that offer promising precedents for others. There are great challenges for city planners, policy makers and developers in cities of developing countries. Addressing the issue of cooler cities requires action across sectors and stakeholders in order to achieve comprehensive and integrated strategies.

A. Cheshmehzangi (✉)
The University of Nottingham Ningbo China, Ningbo, China

C. Butters
University of Warwick, Coventry, UK

A. Cheshmehzangi, C. Butters (eds.), *Designing Cooler Cities*,
Palgrave Series in Asia and Pacific Studies,
https://doi.org/10.1007/978-981-10-6638-2_13

Solutions exist, and many of them are not significantly more costly. Whereas traditional principles for 'staying cool' in hot climates have long been known, modern science, data and research are enabling designers to develop a range of new solutions, both at micro, meso and macro levels. Not least, our knowledge of climatology and of human comfort enable us to understand human needs for healthy urban environments. In view of global warming, growing urban heat production and urban heat island effects, this question is of increasing urgency for millions, especially lower income groups, and especially in the large cities of regions such as Asia.

Broader Relevance

However as noted, this book is not only relevant to Asia, nor only to hot-climate developing-country cities alone. In the rich countries, and even in cold climates, city cooling in summer is, in many cases, already a major issue too, with increasing heat island effects and extreme heat events. And there are other hot-climate world regions that are as yet less developed than many of the Asian economies. In those regions, such as in Africa and Latin America, there are as yet fewer large cities; and in the poorer cities there is as yet less vehicle traffic, air conditioning and other urban heat sources. But many of these regions are likely to follow the same trends as in Asia: towards more compact and higher-rise cities, with less and less green space, more vehicles, more heat sources, and deteriorating living conditions.

Such regions thus have the opportunity to look ahead and make choices of direction for their cities that will have major consequences in the future, both for urban comfort as well as overall costs, energy use and climate emissions. The potential savings – *avoided future impacts*—of preventive action are without doubt very large. Unhappily, in the poorer cities, it is often difficult—if not impossible—to make long-term, sustainability focused plans for urban development; due to a lack of skills, resources, governance, political will or many other pragmatic factors.

From the Local to the Global

It is not difficult to design low energy, low emission and low-cost buildings, in hot and in cold climates. Solutions have existed for many years. The first nearly zero energy buildings were built and tested over 30 years ago. The same applies to urban solutions. Vernacular and traditional examples indicate that good approaches existed long ago, even without access to the modern technologies and methods that are available now. This implies an architecture in evolution, as opposed to one where technology has tended to replace natural principles and, in many cases, common sense. Local solutions for cooler cities improve quality of life at the same time as they mitigate the global issues of rising energy use, pollution and global climate change.

We venture to add that the climate agenda sometimes appears to be a diversion, with many reports and global conferences but little action. Over the years, 'sustainability' has become a very diffuse, misused and one might well say polluted, concept. In our view, most of the advisable actions for pleasant, cool, low-carbon cities would be good policy and good common sense, *with or without a climate problem*. Such qualities as traffic free streets, less traffic noise, less air pollution, better indoor environments, less deforestation, more green space, more recycling, healthy building products and low or even zero energy bills for consumers, are all good solutions regardless of global climate concerns.

Technology and Behaviour

The technology part of the task is, if anything, the easier part. What is more difficult is delivery in the real world; especially at the rate that is needed in order to improve both the living quality of urban millions in the fast-expanding cities of hot climate developing countries, and to safeguard our global climate.

We do not address the equally important behavioural and lifestyle issues but these must be brought into play for a sustainable future. There is growing awareness—with research findings to support it—that the typical technological focus in research and policy is insufficient and even misleading. The performance in real life of low-energy buildings is often far less than was calculated in theory. This puts policy leaders in the position of making unrealistic promises, and building owners in the position of having far longer payback times than expected. Reasons for this are human error, inefficient management, incorrect installation and above all poor user awareness and behaviour. If this is the reality of people, it highlights the inadequacy of theoretical technical efficiency. These 'soft' barriers are culturally specific. The way forward cannot rely on technological innovations alone; planning processes, social awareness and behavioural issues are recognised as being equally important for results. Similarly, key reasons why innovations spread so slowly are not technical or even economic, but human factors such as weak communication, conflicts of interest, poorly designed incentives and policies, as well as consumer fashions that lead us in the opposite direction of unsustainable consumption.

The Dynamics of Change

If the solutions for cooler cities, and more sustainable cities in general, exist, why is this not spreading? The difficulty is to *deliver* them, in the real world. There are many barriers, often cultural and institutional. Delivery—not technological innovation—is the real challenge.

What policy instruments, legislative, economic and cultural, work in different countries? What can be learned from regions where energy efficiency has been successfully introduced and spread? What are the drivers and dynamics of change towards sustainability? Countries are at very different stages of development; opportunities, drivers and barriers vary widely. Examples such as the passive house movement in Europe highlight the local and cultural nature of the contextual dynamics involved. Another relatively successful story is that of building certification schemes such as BREEAM in the UK, LEED in North America, CASBEE in Japan and others in both developed and developing countries. Despite some inherent weaknesses and a certain 'box-ticking' tendency, these have greatly advanced awareness of energy and associated issues in the building industry. Although started slowly, some of them have now become mandatory or are becoming part of national legislation. What can be done to make these processes of change more rapid?

The four main potential agents of change in this field are government, the construction industry, building and design experts, and the public. All four are essential. The initial drivers behind green design were pioneers in North America and Europe—a bottom-up, voluntary process amongst designers and a small public. But green solutions remained very much a minority interest until the goal of sustainable development arose following the 1987 Brundtland Report and the 1992 Rio conference. Thereafter, energy, and in the 1990s climate change, came into focus. In this phase, the third of the above four agents, namely the authorities and public policy, come into play. Coming slowly into play now is fourth of the above agents—the market itself. Business and the construction industry are primarily concerned with profitability, but there are increasing win-win opportunities in green business. The broader goal, which includes corporate social responsibility, is—in the US expression—the 'triple bottom line' of *people, planet, prosperity*. However, experience in Europe shows clearly that the market for sustainable solutions does not emerge by itself, without sufficient science and public support as well as strong public policies. This complex field encompasses perceptions, technologies, market forces, policies and cultural dynamics of many kinds.

Sustainable development and good planning are difficult anywhere; in many developing countries, planning and governance capacity are weak or absent. Alongside gradual capacity building, only quite pragmatic approaches, attuned to local context, can succeed. Sustainable solutions are available, but success is a question of quite long processes. European experience has shown that there are win-win opportunities where environmental and social ambitions can be raised whilst maintaining the 'bottom line' of profitability. Green building is often hardly more expensive once established—though incentives are needed to achieve initial market penetration—and it is good for everyone's pockets, both individual and public finances. There is also a great dynamic in community processes of participation. Sustainable building and urban development almost everywhere has identified and pursued four difficult but essential processes—summed up as follows:

- from segregated spatial zoning of cities to mixed-use districts;
- from specialisation to integrated design and planning, also a key to lower costs;
- from uncontrolled construction to voluntary guidelines to mandatory standards and codes for environmental quality;
- from private-public contradictions to a win-win modus with better cooperation.

All of the above have been the subject of very major efforts and important shifts in policy, planning and practice in industrialised countries.

A Few Key Recommendations

To sum up some of the above remarks and issues highlighted in this book, we state a few key recommendations. These again are not only relevant to the context of Asia, but also to cities in other hot regions of the world.

Compact City Planning

As regards types of city, compact low-dense cities offer environmental, climatic and social advantages compared to high-rise cities, making it easier to provide cool environments, yet can also accommodate fairly high population densities. Cluster-type layouts and compact urban patterns can better utilise passive cooling strategies than large scale urban blocks. Compact layout does not necessarily require a high-rise typology.

Green-Blue Infrastructures

Overall urban planning of green-blue infrastructures is a key to an improved local microclimate, thus improving both outdoor conditions and building cooling loads, as well as reducing energy needs. It is important to consider both *availability* of green spaces and their *accessibility* for residents. Higher priority must be given to open space, green landscaping, vegetation, water bodies and wind corridors in future policies and planning in hot-climate cities.

Climate-Responsive Design

Buildings as well as city layouts should make maximum use of climate-responsive, passive-cooling design; adding the most efficient renewable energy technologies for cooling where needed. This requires a mindset that is aware of those basic principles of regional 'design with nature'. In addition to cultural or aesthetic considerations there are many scientific and climatic grounds for this. Climate-responsive strategies evidenced in traditional hot climate design can often be applied in modernised versions.

Urban and District Scale Cooling Systems

District cooling (DC), like district heating, is a proven technology; but DC is still only occasionally applied in new urban areas of the many hot-climate cities in Asia or elsewhere. There should be a major move towards urban and district scale cooling systems, both for higher efficiencies and especially to remove urban heat and mitigate the urban heat island effect.

Low Carbon Cities

In addition to the buildings themselves, the civil works and infrastructures required in compact high-rise cities make up a large part of the total energy and carbon footprint of such cities, thus adding significantly to global warming; this aspect has been little discussed to date. Whilst dense high-rise may be appropriate for central business districts, it renders the task of low-carbon design more challenging. Embodied carbon, whilst only an indirect contributor to warming cities, requires increased attention as cities become our major form of habitat.

Integration Across Disciplines and Levels

There is great need for integration between the three levels of *building* design, *urban* design and *energy systems* design, as well as between the specialists within the three levels discussed. Lack of integration quickly leads to suboptimal solutions, as well as to larger disparities between various stakeholders and implementation barriers to sustainable development.

Long-Term Planning

Cities exist for centuries. A long-term planning view is ever more essential. The balancing of *pace* and *scale* of development is essential for rapidly growing cities; short-term solutions create negative impacts on city environment and well-being and imply large future costs for energy and health. Rising temperatures in the cities of the hot-climate developing countries is one of the major challenges these cities face.

Index

Note: Page numbers followed by "n" refers to note.

A. Cheshmehzangi, C. Butters (eds.), *Designing Cooler Cities*, Palgrave Series in Asia and Pacific Studies,
https://doi.org/10.1007/978-981-10-6638-2

FSC
www.fsc.org
MIX
Papier | Fördert
gute Waldnutzung
FSC® C083411

Kolibri 360 GmbH
Ferdinand-Jühlke-Straße 7
99095 Erfurt, Deutschland
produktsicherheit@kolibri360.de